Multivariate
Data Analysis
With Readings

THE PPC MARKETING SERIES

Louis E. Boone, Consulting Editor

Louis E. Boone, The University of Tulsa
CLASSICS IN CONSUMER BEHAVIOR

Louis E. Boone, The University of Tulsa
Donald D. Bowen, The University of Tulsa
**THE GREAT WRITINGS IN MANAGEMENT AND
ORGANIZATIONAL BEHAVIOR**

Louis E. Boone, The University of Tulsa
James C. Johnson, St. Cloud State University
MARKETING CHANNELS, Second Edition

Louis E. Boone, The University of Tulsa
David L. Kurtz, Eastern Michigan University
Joseph L. Braden, Eastern Michigan University
THE SALES MANAGEMENT GAME, Second Edition

Joseph F. Hair, Jr., Louisiana State University
Rolph E. Anderson, Drexel University
Ronald L. Tatham, Burke Marketing Research
Bernie J. Grablowsky, Old Dominion University
MULTIVARIATE DATA ANALYSIS WITH READINGS

James C. Johnson, St. Cloud State University
**READINGS IN CONTEMPORARY PHYSICAL DISTRIBUTION,
Third Edition**

James C. Johnson, St. Cloud State University
Donald F. Wood, San Francisco State University
CONTEMPORARY PHYSICAL DISTRIBUTION

Stephen K. Keiser, University of Delaware
Max E. Lupul, California State University, Northridge
MARKETING INTERACTION: A DECISION GAME

Howard A. Thompson, Eastern Kentucky University
THE GREAT WRITINGS IN MARKETING

Multivariate Data Analysis
With Readings

Joseph F. Hair, Jr.
Louisiana State University
Rolph E. Anderson
Drexel University
Ronald L. Tatham
Burke Marketing Research
Bernie J. Grablowsky
Old Dominion University

Division of
Petroleum Publishing Company
Tulsa, Oklahoma 74101

To *Dale*, *Carol*, *Helene* and our *parents*.

Manufactured in the United States of America.

Library of Congress Catalog Card Number 78–532–69

ISBN 0–87814–077–8
1 2 3 4 5   83 82 81 80 79

Cover and Text Design
George J. Alexandres

Written for the non-specialist in multivariate statistics, *Multivariate Data Analysis With Readings* provides a simple introduction to multivariate analysis with emphasis on the practical use of these valuable tools. We specifically wrote this book for those who want a conceptual understanding of multivariate techniques, what they can do . . . when they should be used. . . and how they are interpreted—without becoming bogged down by symbols, formulas, or mathematical derivations. We believe it is the most practical guide available to understanding and applying these otherwise complex statistical tools.

Several statistics texts attempt to explain the fundamentals of multivariate analysis, but most are painfully difficult to read since they assume too much quantitative background on the part of the reader. Generally, these books assume a thorough knowledge of statistics, matrix algebra, and even calculus. Even when students, researchers and managers do have sufficient mathematical training (and most do not), they tend to become so bogged down in the mechanical manipulation of symbols, numbers, and equations that they lose sight of the foundation and usefulness of multivariate statistics. Readers may learn how to substitute numbers for symbols and solve complex equations, yet have little or no understanding of the underlying logic. More important, they do not gain an appreciation for the practical application of multivariate techniques.

You need not be an auto mechanic or mechanical engineer to drive a sophisticated car, nor do you have to be a statistician to use multivariate statistical techniques to analyze data with a computer. Instead of teaching people how to manipulate or crunch numbers, this book helps scholars and managers analyze data more effectively and efficiently by showing how to use statistical methods which allow the study of people and things within the complexity of their natural settings.

This book is intended as a basic text for an introductory course in multivariate analysis. It also may be used as a supplementary text for a research course, a models course, or a second statistics course. Its objective is to provide a fundamental understanding of the nature, power, and the limitations of multivariate statistical techniques for those with a limited statistics and/or mathematics background. Those with a more advanced background will find the book useful for review and reference.

In this book we "talk" our way through the fundamentals upon which the family of multivariate statistical techniques are based. We try to avoid using jargon, stilted phrases, technical "buzz words," and confusing terminology. When necessary, we carefully define any unfamiliar terms introduced to help clarify explanations of the multivariate tools. This approach enables you to grasp the essentials of multivariate analysis and prepares you to make the transition from this introductory book to more quantitative treatments.

This text is organized into three broad areas. The first is presented in Chapter One and consists of an introduction to the field of multivariate data analysis. Each of the multivariate techniques covered in the text is defined, and guidelines for selecting the appropriate application of a technique are provided. Also, a single data bank to be utilized throughout the text for expository purposes is introduced.

The second and third areas of the text are based upon a classification of dependence and interdependence techniques. Chapters two, three, four, and five focus on dependence techniques in which the variables can be divided into independent and dependent classifications. These techniques include: multiple regression analysis, multiple discriminant analysis, multivariate analysis of variance, and canonical correlation analysis. The last two chapters deal with the analysis of interdependence in which the variables or subsets of variables are not categorized as either independent or dependent. Rather, the entire set of variables is dealt with as a group, and an effort is made to give meaning to the set of variables, individuals, or objects. The techniques included here are factor analysis, multidimensional scaling and conjoint analysis.

Each chapter of the book consists of a text portion followed by articles carefully selected from the behavioral literature. The text portion introduces you to the fundamentals of the technique and enables you to understand the basic issues surrounding the application of the technique. Following the text portion are articles which set forth the theory and illustrate it in a specific application. We believe that it is both instructive and reassuring for you to see how others have applied the techniques.

We wrote this book for people who want an overall view of what multivariate techniques can do for them, when and how they can be used, and how results are interpreted so that they can use the tools successfully and read the more technical literature with confidence. We believe we have achieved these objectives!

Our heartfelt appreciation goes to our families and friends for their support and encouragement, and to the many writers and teachers from whom we learned. We are especially grateful and want to give special thanks to Dr. Walter A. Smith of Tulsa University, who read through the entire final manuscript of this book. Many of his suggestions are incorporated herein. In particular, we want to thank those who by asking at least once a day, "When will the book be finished?," goaded us into completing it. Any errors in this book we pass on to writing style, not ignorance.

Joseph F. Hair, Jr., *Louisiana State University*
Rolph E. Anderson, *Drexel University*
Ronald L. Tatham, *Burke Marketing Research*
Bernie J. Grablowsky, *Old Dominion University*

Table of Contents

1

An Introduction to Multivariate Analysis

CHAPTER REVIEW

This chapter presents a simplified introduction to the rationale of multivariate data analysis. It notes that multivariate analysis methods will increasingly influence not only the analytical aspects of research, but also the design and approaches to data collection for decision-making/problem-solving situations. A classification of the various types of multivariate techniques is presented and general guidelines for their application are provided. The chapter concludes with a discussion of the data bank which is utilized throughout the text to illustrate the application of each of the techniques. Before beginning the chapter you should familiarize yourself with the definitions of key terms to facilitate your reading.

After studying the concepts presented in this introductory chapter, you should be able to do the following:

- [ ] Understand what multivariate analysis is and when its application is appropriate.
- [ ] Define and discuss the specific techniques included in multivariate analysis.
- [ ] Determine which multivariate technique is appropriate for a specific research problem.
- [ ] Discuss the nature of measurement scales and their relationship to multivariate techniques.

DEFINITIONS OF KEY TERMS

DEPENDENCE TECHNIQUE.   A classification of statistical techniques distinguished by having a variable or set of variables identified as the dependent variable and the remaining variables as independent. An example is regression analysis.

INTERDEPENDENCE TECHNIQUE.   A classification of statistical techniques for which the variables are not divided into dependent and independent groups. Rather, all variables are analyzed as a single set. An example is factor analysis.

NONMETRIC DATA.   Also referred to as qualitative data, they are attributes, characteristics or categorical properties that can be used to identify or describe a subject or object. Examples of nonmetric data are sex (male, female) and occupation (physician, attorney, professor).

METRIC DATA.   Also referred to as quantitative data, they are measurements used to identify or describe subjects (or objects) not only on the basis of type or kind, but also as to the amount or degree to which the subject may be characterized by a particular attribute. For example, a person's age and weight are considered metric data.

DUMMY VARIABLE. A nonmetrically measured variable which has been transformed into a metric dummy variable by assigning a "1" or a "0" to a subject depending upon whether it possesses or does not possess a particular characteristic.

INDEPENDENT VARIABLE. The presumed cause of any change in a response or dependent variable.

DEPENDENT VARIABLE. The presumed effect of, or response to a change in an independent variable.

TREATMENT. The independent variable that the researcher manipulates to see the effect (if any) on the dependent variables.

UNIVARIATE ANALYSIS OF VARIANCE (ANOVA). A statistical technique used to determine, on the basis of one dependent measure, whether samples are from populations with equal means.

BIVARIATE PARTIAL CORRELATION. Simple (two-variable) correlations between the two sets of residuals (unexplained variances) that remain after the association of other independent variables is removed.

1

AN INTRODUCTION TO MULTIVARIATE ANALYSIS

In recent years remarkable advances have been made in the analysis of psychological, sociological and other types of behavioral data. Computer technology has made it possible to analyze large quantities of complex data with relative ease. At the same time, the ability to conceptualize data analysis has advanced too, although perhaps not as rapidly as computer technology. Much of the increased understanding and mastery of data analysis has come about through the study of statistics and statistical inference. Equally important has been the expanded understanding and application of a group of analytical statistical techniques known as multivariate analysis.

Multivariate analytical techniques are beginning to be applied on a widespread basis in industry, government and university-related research centers. Several books and articles have been published on the theoretical and mathematical aspects of these tools. Few books, however, have been written for the researcher who is not a specialist in math or statistics. Fewer such books discuss applications of multivariate statistical methods of interest to behavioral scientists, or business and government managers who want to expand their knowledge of multivariate analysis to better understand increasingly complex phenomena in their work environment. For example, the businessperson is interested in learning how to develop strategies to appeal to customers with varied demographic and psychographic characteristics in a marketplace with multiple constraints (legal, economic, competitive, technological, etc.). Clearly, multivariate techniques are required to adequately study these multiple relationships and obtain a complete, realistic understanding for decision-making.

Any researcher who examines only two variable relationships when investigating phenomena under conditions approaching natural complexity either assumes that ignoring other variables has no effect on results, or that there are no better tools for examining more complex relationships. As one researcher states: "For the purposes of . . . any . . . applied field, most of our tools are, or should be, multivariate. One is pushed to a conclusion that unless a . . . problem is treated as a multivariate problem, it is treated superficially [7]. According to statisticians Hardyck and Petrinovich [8]:

"... multivariate analysis methods will predominate in the future and will result in drastic changes in the manner in which research workers think about problems and how they

4

design their research. These methods make it possible to ask specific and precise questions of considerable complexity in natural settings. This makes it possible to conduct theoretically significant research and to evaluate the effects of naturally occurring parametric variations in the context in which they normally occur. In this way, the natural correlations among the manifold influences on behavior can be preserved and the separate effects of these influences can be studied statistically without causing a typical isolation of either individuals or variables."

Application of the computer to process large, complex data banks has spurred the use of multivariate methods. Today, a number of "canned" computer programs are available for data analysis, and others are being developed [2, 5, 6, 9, 11]. In fact, many researchers have appeared who realistically call themselves "data analysts," instead of mathematicians or, in the vernacular, "quantitative types." These data analysts have contributed substantially to the increase in the number of journal articles using multivariate statistical techniques. Even for people with strong quantitative training, the availability of canned programs for multivariate analysis has facilitated the complex manipulation of data matrices that have long hampered the growth of multivariate techniques.

WHAT IS MULTIVARIATE ANALYSIS?

Multivariate analysis is difficult to define. Broadly speaking, it refers to all statistical methods which simultaneously analyze multiple measurements. Any simultaneous analysis of more than two variables can be loosely considered multivariate analysis. As such, multivariate techniques are extensions of univariate analysis (analysis of single variable distributions), and bivariate analysis (cross-classification, correlation, and simple regression used to analyze two variables).

One of the reasons for the difficulty in defining multivariate analysis is that the term "multivariate" is not used consistently in the literature. To some researchers "multivariate" simply means examining relationships between or among more than two variables. Others use the term only for problems where there are multiple variables all of which are assumed to have a multivariate normal distribution. To be considered truly multivariate, all the variables must be random variables which are interrelated in such ways that their different effects cannot easily be studied separately. Additionally, some authors purport that the purpose of multivariate analysis is to measure, explain, and/or predict the degree of relationship among variates (weighted combinations of variables). Thus, the multivariate character lies in the multiple variates (multiple combinations of variables), not solely in the number of variables or observations [7]. For the purpose of this text we prefer not to follow a strict definition of multivariate analysis. Instead, both multi-variable techniques and truly multivariate techniques will be discussed. This approach is followed because the authors believe that knowledge of multi-variable techniques is essential to understanding multivariate analysis.

TYPES OF MULTIVARIATE TECHNIQUES

Multivariate analysis is a relatively new but rapidly expanding approach to data analysis. Specific techniques included in multivariate analysis are (1) multiple regression and multiple correlation; (2) multiple discriminant analysis; (3) principle components analysis and common factor analysis; (4) multivariate analysis of variance and covariance; (5) canonical correlation analysis; (6) cluster analysis; (7) multi-dimensional scaling and (8) conjoint analysis. The first four categories of techniques are frequently used by practicing statisticians, and all but the fourth are fairly well established in academic research literature. The four remaining techniques are less well known and have been applied only tentatively and experimentally by researchers. A separate chapter in this text is devoted to each of these techniques except cluster analysis and conjoint analysis, which are discussed as extensions of the other techniques. At this point we will introduce the reader to each of the multivariate techniques by briefly defining the technique and the objective for its application.

Multiple Regression (MR) is the method of analysis which is appropriate when the research problem involves a single, metric dependent variable presumed to be related to one or more metric independent variables. The objective of multiple regression analysis is to predict the changes in the dependent variable in response to changes in the several independent variables. This objective is most often achieved through the statistical rule of least squares.

Whenever the researcher is interested in predicting the level of the dependent variable, he will find multiple regression useful. For example, monthly expenditures on dining out (dependent variable) might be predicted from information regarding a family's income, the size of the family, and the age of the head of household (independent variables). Similarly, the researcher might attempt to predict a company's sales from information on the company's expenditures for advertising, the number of salespeople, and the number of stores selling its products.

If the single dependent variable is dichotomous (e.g., male-female) or multichotomous (e.g., high-medium-low) and therefore nonmetric, the multivariate technique of *Multiple Discriminant Analysis* (MDA) is appropriate. Discriminant analysis is useful in situations where the total sample can be divided into groups based on a dependent variable which has several known classes. The primary objectives of multiple discriminant analysis are to understand group differences and predict the likelihood that an entity (individual or object) will belong to a particular class or group based on several metric independent variables. For example, discriminant analysis might be used to distinguish innovators from non-innovators according to their demographic and psychographic profiles. Other applications include distinguishing heavy product users from light users, males from females, national brand buyers vs. private label buyers, and good credit risks from poor credit risks.

Multivariate Analysis of Variance (MANOVA) is a statistical technique which can be used to explore simultaneously the relationship between several categorical independent variables (usually referred to as treatments) and two or more metric dependent variables. As such, it represents an extension of univariate analysis of variance (ANOVA). Multivariate analysis of covariance (MANCOVA) can be used in conjunction with MANOVA to remove (after the experiment) the effect of any uncontrolled independent variables on the dependent variables. The procedure is similar to that involved in bivariate partial correlation. MANOVA is useful when the researcher designs an experimental situation (manipulation of several non-metric treatment variables) to test hypotheses concerning the variance in group responses on two or more metric dependent variables.

Canonical Correlation Analysis (CCA) can be viewed as a logical extension of multiple regression analysis. Recall that multiple regression analysis involves a single metric dependent variable and several metric independent variables. With canonical analysis the objective is to simultaneously correlate several metric dependent variables and several metric independent variables. Whereas multiple regression involves a single dependent variable, canonical correlation involves multiple dependent variables. The underlying principle is to develop a linear combination of each set of variables (both independent and dependent) in a manner which maximizes the correlation between the two sets. Stated in a different manner, the procedure involves obtaining a set of weights for the dependent and independent variables which provide the maximum simple correlation between the set of dependent variables and the set of independent variables.

The techniques discussed thus far have focused on multivariate methods applied to data which contain both dependent and independent variables. However, if the researcher is investigating the interrelations, and therefore the interdependence among all the variables, without regard to whether they are dependent or independent variables, then several other multivariate methods are appropriate. These methods include factor analysis, cluster analysis, multidimensional scaling, and conjoint analysis.

Factor Analysis (FA), including its variations such as component analysis and common factor analysis, is a statistical approach which can be used to analyze the interrelationships among a large number of variables and then explain these variables in terms of their common, underlying dimensions (factors). The statistical approach involves finding a way of condensing the information contained in a number of original variables into a smaller set of dimensions (factors) with a minimum loss of information.

Cluster analysis is an analytical technique which can be used to develop meaningful subgroups of individuals or objects. Specifically, the objective is to classify a sample of entities (individuals or objects) into a small number of mutually exclusive groups based on the similarities among the entities. Unlike discriminant analysis the groups are not pre-defined. Instead, the technique is used to identify the groups.

Cluster analysis usually involves at least two separate steps. The first is the measurement of some form of similarity or association between the entities in order to determine how many groups really exist in the sample. The second step is to profile the persons or variables in order to determine their composition. This second step may be accomplished by applying discriminant analysis to the groups identified by the cluster technique.

In *Multidimensional Scaling* (MDS), the objective is to transform consumer judgements of similarity or preference (e.g., preference between stores or brands) into distances represented in a multidimensional space. If objects A and B are judged by the respondents as being most similar compared to all other possible pairs of objects, multidimensional scaling techniques will position objects A and B in such a way that the distance between them in multidimensional space is shorter than that between any other two objects. A related scaling technique, *Conjoint Analysis,* is concerned with the joint effect of two or more independent variables on the ordering of a single dependent variable. It permits development of stronger measurement scales by transforming rank order responses into metric effects. As will be seen, metric and nonmetric multidimensional scaling techniques produce similar appearing results. The fundamental difference is that metric multidimensional scaling uses metric data to position the objects, whereas nonmetric multidimensional scaling involves a preliminary transformation of the nonmetric data into metric data, and then the objects are positioned using the transformed data.

A CLASSIFICATION OF MULTIVARIATE TECHNIQUES

The broad concept of multivariate analysis and some of the more recently developed specific techniques that belong with it may be relatively new to the reader. To assist you in becoming familiar with these techniques, a classification of most of the multivariate methods is presented in Figure 1.1. This classification is based on three judgments the analyst must make about the nature and utilization of the data: (1) Can the variables be divided into independent and dependent classifications based on some theory? (2) If yes, how many variables are treated as dependent in a single analysis? (3) How are the variables measured? The selection of the appropriate multivariate technique to be utilized will depend upon the answers to these three questions.

In this chapter and throughout the book, there is a close relationship between the various multivariate methods, and they can be viewed as a family of techniques. When considering the application of multivariate statistical techniques the first question to be asked is: "Can the data variables be divided into independent and dependent classifications?" The answer to this question will indicate whether a "dependence" or "interdependence" technique should be utilized. Note that in the classification shown in Figure 1.1, the dependence techniques are on the left side and the interdependence techniques are on the right. A *dependence technique* may be defined as one in which a variable or set of variables is identified as the dependent variable to be predicted by or explained by other independent variables. An example of a dependence technique is multiple regression analysis. In contrast, an *interdependence technique* is one in which no single variable or group of variables is defined as being independent or dependent. Rather, the procedure involves the analysis of all variables in the set simultaneously. Factor analysis is an example of an

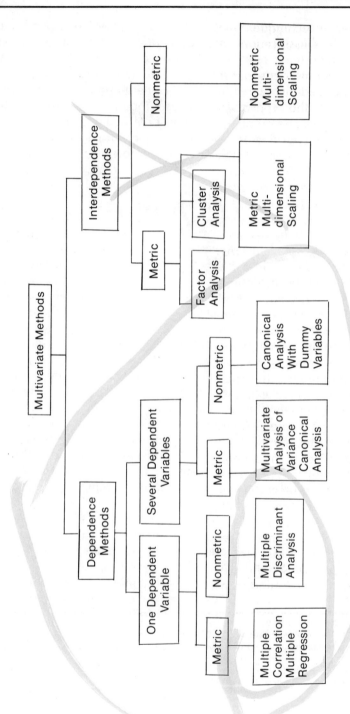

Figure 1.1—A Classification of Multivariate Methods

Source: Adapted from J. Sheth, "*The Multivariate Revolution in Marketing Research*," *Journal of Marketing* (January, 1971).

interdependence technique. Let's focus on dependence techniques first and use the classification in Figure 1.1 to select the appropriate multivariate method.

The different methods that constitute the analysis of dependence can be categorized by two things: (1) the number of dependent variables, and (2) the type of measurement scale employed by the variables. Regarding the number of dependent variables, dependence techniques can be classified as those having either a single dependent variable or those having several dependent variables. Dependence techniques can be further classified as those with either metric (quantitative/numerical) or nonmetric (qualitative/categorical) dependent variables (the concept of metric or nonmetric variables will be discussed in the following section on measurement scales). If the analysis involves a single dependent variable which is metric, then the appropriate technique would be either multiple correlation or multiple regression analysis. On the other hand, if the single dependent variable is nonmetric (categorical), then the appropriate technique would be multiple discriminant analysis. In contrast, when the research problem involves several dependent variables, then two other techniques of analysis are appropriate. If the several dependent variables are metric, then the techniques of multivariate analysis of variance or canonical correlation analysis are appropriate. If the several dependent variables are nonmetric, then they can be transformed through dummy variable coding (0-1) and canonical analysis can be used.[1]

The interdependence techniques are shown on the right side of the classification scheme in Figure 1.1. The reader will recall that with interdependence techniques the variables are not able to be classified as either dependent or independent variables. Instead, all the variables are analyzed simultaneously in an effort to give meaning to the entire set of variables or subjects. As with dependence techniques, the interdependence techniques can also be classified as either metric or nonmetric. Generally, factor analysis and cluster analysis are considered to be metric interdependence techniques. However, nonmetric data may be transformed through dummy variable coding for use with factor analysis and cluster analysis. Both metric and nonmetric approaches to multidimensional scaling have been developed.

---

[1]Dummy variable coding will be discussed in greater detail later. Briefly, dummy variable coding is a means of transforming nonmetric data into metric data. It involves the creation of so called dummy variables, in which 1's and 0's are assigned to subjects depending on whether they possess or do not possess a characteristic in question. For example, if a subject is a male, assign him a 1; if the subject is a female, assign her a 0—or the reverse.

Measurement Scales

The basic nature of most data analysis involves the partitioning, identification and measurement of variation in a dependent variable due to one or more different independent variables. The key word here is measurement because the researcher cannot partition or identify variation unless it can be measured. Measurement is important not only in data analysis, but also in the selection of the appropriate multivariate method of analysis. In the next few paragraphs we shall discuss the concept of measurement as it relates to data analysis and particularly its relationship to the various multivariate techniques.

There are two basic kinds of data: nonmetric (qualitative) and metric (quantitative). Nonmetric data are attributes, characteristics, or categorical properties that can be used to identify or describe a subject. While nonmetric data differ in type or kind, metric data measurements are made so that subjects may be identified as differing in amount or degree. Metrically measured variables reflect relative quantity or distance, whereas nonmetrically measured variables do not. As seen in Figure 1.2, nonmetric data are measured with nominal or ordinal scales and metric variables are measured using interval and ratio scales.

Measurement with a nominal scale involves assigning numbers which are used to label or identify subjects or objects. Nominal scales

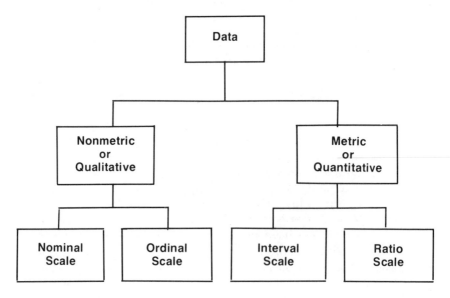

Figure 1.2—Types of Data and Measurement Scales

provide the least precise measurement since the data consists merely of the number of occurrences in each class or category of the variable being studied. Therefore, the numbers or symbols assigned to the objects have no quantitative meaning beyond indicating the presence or absence of the attribute or characteristic under investigation. Examples of nominally scaled data include an individual's sex, religion, or political party. In working with these data, the analyst might assign numbers to each category, e.g., 1's for females and 0's for males. The numbers here only represent categories or classes and do not imply amounts of an attribute or characteristic.

Ordinal scales are the next higher level of measurement precision. Variables can be ordered or ranked with ordinal scales in relation to the amount of the attribute possessed. Every subclass can be compared with another in terms of a "greater than," or "less than" relationship. For example, different levels of an individual consumer's satisfaction with several new products can be illustrated on an ordinal scale. The following scale shows a respondent's view of three products. The respondent is more satisfied with A than B, and more satisfied with B than C.

Very Satisfied	Product A	Product B	Product C	Very Unsatisfied

Numbers utilized in ordinal scales such as the above are nonquantitative, since they indicate only relative positions in an ordered series. There is no measure of how much satisfaction the consumer receives in absolute terms, nor does the researcher know the exact difference between points on the scale of satisfaction. Most scales in the behavioral sciences fall into this ordinal category.

Interval scales and ratio scales (both metric) provide the highest level of measurement precision. Thus, they permit nearly all mathematical operations to be performed. These two scales have constant units of measurement, so differences between any two adjacent points on any part of the scales are equal. The only real difference between interval and ratio scales is that interval scales have an arbitrary zero point, while ratio scales have an absolute zero point. The most familiar interval scales are the Fahrenheit and Centigrade temperature scales. Both have a different arbitrary zero point, and neither really indicates a zero amount or lack of temperature, since we can register temperatures below the zero point of each scale. Therefore it is not possible to say that any value on an interval scale is some multiple of some other point on the scale. For example, an 80° day cannot correctly be said to be twice as hot as a 40° day.

Ratio scales represent the highest form of measurement precision since they possess the advantages of all lower type scales plus an absolute zero point. All mathematical operations are allowable with ratio scale measurements. The bathroom scale or other common weighing machines are examples of these scales, for they have an absolute zero point and one can speak in terms of multiples when relating one point on the scale to another, i.e., 100 pounds is twice as heavy as 50 pounds.

Understanding the different types of measurement scales is important in determining which multivariate technique is most applicable for the data. Recall that two of the considerations in developing the classification of multivariate methods presented in Figure 1.1 were how the variables are measured and whether they can be divided into independent and dependent classifications. These two considerations are summarized in matrix form in Figure 1.3. The matrix can be used to select the appropriate multivariate technique by variable and data types. For example, if the independent variables are metrically scaled

Dependent Variables

		Metric	Nonmetric
Independent Variables	Metric	Multiple Regression Analysis Canonical Correlation Analysis	Multiple Discriminant Analysis Canonical Correlation With Dummy Variables
	Nonmetric	Multivariate Analysis of Variance	Canonical Correlation With Dummy Variables

Figure 1.3—Selection Of Multivariate Techniques *by Data and Variable Types*

and the dependent variable(s) is nonmetric, then the appropriate techniques are multiple discriminant analysis and/or canonical correlation with dummy variables. Thus, by identifying the variable type (independent or dependent) and the data type (metric or nonmetric), one can select the appropriate multivariate technique to utilize in analyzing the data.

DATA BANK

To explain and illustrate each of the multivariate techniques, a single data bank is utilized throughout the book. The data bank is a hypothetical set of data generated by the authors to meet the assumptions required in the application of multivariate techniques. The data bank consists of 50 observations on ten separate variables. We will assume that the data were obtained from the Hair, Anderson, Tatham Company (HATCO). HATCO is interested in identifying (and hopefully predicting) which individuals will be knowledgeable and highly motivated salespeople. The data provided in the data bank will enable HATCO to develop a profile of the type of individual most likely to be knowledgeable and successful, and also will make it possible to evaluate two different approaches to training. A brief description of the data base variables is provided in Table 1.1, in which the variables also are classified as to whether they are independent or dependent, and metric or nonmetric. A listing of the data bank is provided in the

TABLE 1.1
DESCRIPTION OF DATA BASE VARIABLES

Variable Description	Variable Type
X_1 Self-esteem	Independent/Metric
X_2 Locus of Control	Independent/Metric
X_3 Alienation	Independent/Metric
X_4 Social Responsibility	Independent/Metric
X_5 Machiavellianism	Independent/Metric
X_6 Political Opinion	Independent/Metric
X_7 Knowledge	Dependent/Metric
X_8 Motivation	Dependent/Metric
X_9 Type of Training —role playing (coded 1) —case study (coded 0)	Independent/Nonmetric
X_{10} Top Salesperson Award	Dependent/Nonmetric

accompanying instructor's manual for those who wish to reproduce the solutions reported in this text.

During the first three months of employment, HATCO obtains measures of social-psychological attitudes from all its employees. These measures represent the first six variables of the data set. The six variables are measures of self esteem, locus of control, alienation, social responsibility, Machiavellianism, and political opinion.

During the past several years HATCO has been testing two different methods of sales training to supplement regular reading assignments and lectures in the training of fifty salespeople. The two different methods of sales training being tested are: (1) role playing and (2) case study analysis. Prior to the training programs, each of the trainees is randomly assigned to one of the two different sales training methods. Thus, in our example we will assume 25 of the trainees were assigned to the role playing situation and 25 to the case study situation. These two methods of training are represented by variable nine— type of training.

At the end of the training program each of the sales trainees is given two tests. One test measures overall knowledge about the company's products, and the other test measures motivation toward achievement in the trainee's sales career. Variables seven and eight represent the results for these tests. HATCO also has been monitoring the trainee's performance following the training period. Company records indicate that 30 of the trainees were awarded the "Top Salesperson of the Month Award" after completion of the training program. Variable 10 records whether or not a sales trainee received the top salesperson award.

Variables seven, eight, nine and ten have already been described in sufficient detail. However, additional information needs to be provided on variables one through six to clarify their interpretation and facilitate understanding their use in multivariate techniques. The following definitions, adapted from Robinson and Shaver [10], may be utilized:

X_1 SELF ESTEEM:
This variable is designed to measure attitudes toward the self along a favorable to unfavorable dimension. For example, high self-esteem means the individual respects himself, considers himself worthy, but does not necessarily consider himself better than others and definitely does not consider himself worse. A person with high self-esteem does not feel he is the ultimate in perfection but, on the contrary, recognizes his limitations

and expects to grow and improve. In contrast a person with low self-esteem would be the opposite of the characteristics describing a person with high self-esteem. This variable is scored so that high values indicate a high self-esteem score.

X_2 LOCUS OF CONTROL:
Concerns the degree to which individuals perceive themselves as being in control of their lives and the events that influence it. Some people tend to attribute the influences on their lives to forces that are within their own control, such as their own work efforts and skills. Yet, some people explain their experiences in life by reference to factors beyond their control, such as luck, chance, or strong forces they cannot overcome. The first group of people have an internal locus of control, while the latter group perceive an external locus of control. This variable is scored so that a high value indicates an external locus of control and a low value an internal locus of control.

X_3 ALIENATION:
Alienation is defined as the subjective feeling of estrangement from society and its culture. The concept of alienation is defined and measured through a number of separate components, including powerlessness, normlessness, meaninglessness, cultural estrangement, social estrangement, and work estrangement. Persons who exhibit alienation believe they can be characterized by these six components. This variable is scored so that a high value indicates an individual who is not alienated, and a low value an individual who tends to be alienated.

X_4 SOCIAL RESPONSIBILITY:
This variable assesses a person's traditional social responsibility and orientation toward helping others, even when there is nothing to be gained from helping them. The results of this variable are likely to be the opposite of alienation. A high score on this variable indicates a person who exhibits higher social responsibility, and vice versa.

X_5 MACHIAVELLIANISM:
This variable attempts to measure a person's general strategy for dealing with people, especially the degree to which they feel other people are manipulable in interpersonal situations. Persons who score high on the Machiavellian scale tend to have a cool

detachment, which makes them less emotionally involved with other people, with sensitive issues, or with saving face in embarrassing situations. Also a highly Machiavellian type individual would tend to be more manipulative and impersonal in dealing with other individuals. A high score indicates a person characterized by high Machiavellianism, and vice versa.

X_6 POLITICAL ORIENTATION:
This variable measures an individual's political orientation, and focuses particularly on a continuum of opinions from radicalism to conservatism. It consists of a number of different measures relating to topics of religion, welfare, unionism, government, morals, racial tolerance, contraception, and the legal system. A high score on this variable indicates that an individual tends to be more conservative, and a low score indicates a more liberal individual.

As noted earlier, these data bank variables will be utilized throughout the text when examples of actual results of computer programs are discussed. Two exceptions to this should be noted. In some instances other hypothetical problems for HATCO are used as examples, and in Chapter 7: Multidimensional Scaling and Conjoint Analysis, a unique set of examples will be provided.

SUMMARY

This chapter has introduced you to the exciting, challenging topic of multivariate data analysis. The following chapters discuss each of the individual techniques in sufficient detail to enable the novice data analyst to understand what a particular technique can achieve, when and how it should be applied, and how the results of its application are to be interpreted. End-of-chapter readings from academic literature further demonstrate the application and interpretation of the techniques.

END OF CHAPTER QUESTIONS

1. What are several factors contributing to the increased application of techniques for multivariate data analysis in recent years?
2. How would you define multivariate analysis?
3. List and define the eight techniques for multivariate data analysis described in this chapter.

4. Explain why and how the various multivariate methods can be viewed as a family of techniques.
5. Why is knowledge of measurement scales important to an understanding of multivariate data analysis?

REFERENCES

1. Anderson, T. W. *An Introduction to Multivariate Analysis*. New York: Wiley, 1958.
2. Barr, A. J. et al (eds.) *A User's Guide to SAS 76*. Raleigh, North Carolina: Sparks Press, 1976.
3. Cattell, R. B. (e.). *Handbook of Multivariate Experimental Psychology*.
4. _____. "Guest Editorial: Multivariate Behavioral Research and the Integrative Challenge," *Multivariate Behavioral Research*. Vol. 1, January, 1966, pp. 4–23.
5. Cooley, W. W. and P. R. Lohnes, *Multivariate Data Analysis* (New York: John Wiley & Sons, 1971.
6. Dixon, W. J., editor, *Biomedical Computer Programs*. Los Angeles: Health Science Computing Faculty, School of Medicine, University of California, 1973.
7. Gatty, R. "Multivariate Analysis for Marketing Research: An Evaluation," *Applied Statistics*, Vol. 15, November, 1966, pp. 157–72.
8. Hardyck, C.D. and L.F. Petrinovich. *Introduction to Statistics for the Behavioral Sciences*, Second Edition. Philadelphia: W. B. Sanders Company, 1976.
9. Nie, N. H., C. Hull, J. Jenkins, K. Steinbrenner and D. Bent. *Statistical Package for the Social Sciences*, New York: McGraw-Hill Co., 2nd Edition, 1975.
10. Robinson, J. P., and P. R. Shaver, *Measures of Social Psychological Attitudes*, Survey Research Center, Institute of Social Research, The University of Michigan, Ann Arbor, Michigan, Revised Edition, 1973.
11. Veldman, D. J. *Fortran Programming for the Behavioral Sciences*, New York: Holt, Rinehart, Winston, Inc., 1967.

SELECTED READINGS

HOW TO GET THE MOST OUT OF MULTIVARIATE METHODS
JAGDISH N. SHETH

Today multivariate methods are widely used (and often misused) by many marketing professionals and academic researchers.

Reprinted by permission of Jagdish N. Sheth and the publisher European Research *(November 1976) pp. 229–235.*

Multivariate methods should or actually have practically replaced the more traditional statistical analyses such as frequency distributions and cross-tabulations in marketing research.

It is, therefore, almost impossible today to come across a research study reported in academic journals such as *Journal of Marketing Research* which does not utilize some type of multivariate analysis of the data.

If multivariate analysis is somehow missing in the study, it is often recognized as a weakness to be rectified in a follow-up research. While this widespread use of multivariate methods has certainly increased the respectability of marketing as a discipline among the more 'scientific' and traditional disciplines, it has also brought upon many the problems of communication and understanding. Probably a majority of JMR readers have difficulties comprehending the articles published in it. This problem seems more vivid between the academic and the professional researchers and especially between the researchers and the managers of the marketing function in the organization. Accordingly, there are two objectives of this paper:

1. Provide a nonstatistical description of the multivariate methods and discuss their potential to solve marketing problems.
2. Provide guidelines to the researcher for getting more out of the multivariate methods.

Description and applications of multivariate methods

Multivariate methods refer to those statistical techniques which focus upon, and bring out in bold relief, the structure of *simultaneous* relationships among three or more phenomena. It is important to remember that what matters in multivariate analysis is the analysis of the *simultaneous* relationships among phenomena.

Many sequential or hierarchial statistical analyses of a large number of variables such as the Automatic Interaction Detection (AID) are, therefore, not truly multivariate in nature but simply repeated applications of simpler statistical techniques.

Multivariate methods differ from simple univariate (single phenomenon) statistical techniques in terms of a shift in focus away from the levels (averages) and distributions (variances) of the phenomena, and instead concentrating upon the degree of relationships (correlations or covariances) among these phenomena. They also differ from the bivariate (two phenomena) statistical techniques by shifting focus away from pairwise relationships to the more complex simultaneous relationships among phenomena.[5,6]

Multivariate methods can be broadly categorized into two types: Functional and structural multivariate techniques.

Functional Multivariate Methods. The functional multivariate methods are most appropriate for building predictive models with which the researcher can forecast, or explain one or more phenomena from the knowledge of other phenomena based on their relationships. In order to satisfactorily utilize the functional multivariate methods it is essential that the researcher has considerable knowledge or theory about the market behavior with which to properly conceptualize a realistic model. The functional multivariate methods provide to the researcher estimates of both the directionality and the magnitude of relationships among phenomena. As such, they border on being the most precise quantitative models of market behavior. Of course, how realistic these models may be is a direct function of the imagination and the experience of the model builder.

Depending on the nature and the number of phenomena the researcher wishes to predict or explain, there are several different types of functional multivariate methods. The first most commonly known and used multivariate method is multiple regression, and its many variations, which enables the researcher to predict the level or the magnitude of a phenomenon such as the sales volume or market share of a brand. The objective in multiple regression is to search for the best possible (optimum) *simultaneous* relationship between the distribution of the predicted phenomenon and those of the many other correlated or causal phenomena resulting in establishment of a functional relationship between the criterion and the predictor variables. For example, market shares of grocery products may well be a function of a number of marketing mix variables such as average unit price, customer loyalty, media advertising, store coverage, store display and point-of-purchase promotion.[1]

A second functional multivariate method is multiple discriminant analysis and many of its variations, which are extremely useful if the researcher is interested in predicting the likelihood of an event happening sometime in the future. For example, what is the likelihood that a tax payer's return will be audited by the IRS? Or what is the likelihood that a telephone customer will convert to TouchTone service when it is promoted by the company? The objective in multiple discriminant analysis is to identify those key descriptors on which various predefined events have statistically significant differences, and to build a functional model out of them which will enable the researcher to predict likelihoods of events happening as best as possible. Thus, a tax payer with more than $50,000 taxable income or one who belongs to certain occupations such as medical doctors may have significantly higher likelihood of being audited than the average tax payer. Similarly, a household in upper socioeconomic class, younger life cycle or with high mobility may have significantly greater likelihood of becoming a Touch Tone subscriber than the average customer.

A third functional multivariate method is multivariate analysis of variance (MANOVA) which is more useful for testing the impact of various levels of one or more experimental factors on a variety of phenomena. For example, what is the impact of doubling the advertising budget (weight) on market awareness, attitudes, and purchase behavior toward a product? The objective in MANOVA is to test for significant differences on a set or *profile* of variables due to some changes in one or more causal factors. Thus, for example, doubling the advertising budget may be highly effective in significantly increasing the awareness and attitude levels among customers but may have little impact on their immediate purchase behavior.

A fourth major functional multivariate method is canonical correlation analysis which enables the researcher to build a predictive model with which he can simultaneously forecast or explain several phenomena based on his knowledge of their correlates. For example, the researcher may be interested in the nature and magnitude of price competition among various brands of a product class. The objective in canonical correlation analysis is to simultaneously regress a set of criterion variables on a set of predictor variables in the hope of bringing out the functional relationships both within and between the two sets of variables. Thus it is quite possible that price competition may be prevalent within the national brands and within the store brands but not between the two types of brands.

Structural Multivariate Methods. The structural multivariate methods, on the other hand, are more descriptive and less predictive in nature. They are essentially data reduction techniques which simplify complex and diverse relationships among phenomena in a manner which enables the researcher to gain insights into the underlying and nonintuitive structure of relationships. The structural multivariate methods are thus analogous to the search for the needle in the hay stack.

The most popular of the structural multivariate methods is factor analysis and many of its variations. Factor analysis enables the researcher to gain insights into the common underlying bonds or dimensions by which otherwise highly divergent phenomena tend to correlate among themselves. For example, what is the common bond between income, education and occupation of a household? Or is there any systematic pattern of preferences in the viewership of vast variety of television programs? The objective in factor analysis is to decompose into meaningful components or dimensions the extent of relationships empirically observed among a set of divergent phenomena. Thus, the common underlying dimension of social class may be responsible for the strong positive correlations found between income, education and occupation. Similarly, interest in situation comedy, quiz shows, soap operas, westerns, police or detective stories, etc. may be the common bondages among the vast variety of television programs.

A second structural multivariate method is cluster analysis which enables the researcher to classify, segment or disaggregate entities into homogeneous subgroups based on their similarities on a profile of information. For example what are the different psychographic segments of self-medicated drug users; or what are the different benefit segments among bank customers? The objective in cluster analysis is to meaningfully classify a group of entities into clusters based on some judgemental or statistical rule. There are many different algorithms proposed for cluster analysis and very few have any statistical inferential properties so that cluster analysis is more a heuristic than a statistical technique. However, it does provide insights into the typology or segments present in the data. Thus, it is possible to find psychographic segments such as hypochondriacs, skeptics, realists and authority-seekers in the self-medicated drug case [8], and segments such as social interaction-oriented, banking tasks-oriented and money borrowing-oriented customers in the bank services case.

A more recent structural multivariate method is multidimensional scaling which enables the researcher to explore and infer underlying criteria or dimensions that people utilize to form perceptions about similarities between, and preferences among, various products or services. For example, how do people judge similarities among automobiles or toothpastes? The objective in multidimensional scaling is to map the alternatives in a multidimensional space in such a way that their relative positions in the space reflect the degree of perceived similarity between alternatives. In the process, it provides the researcher insights into the complexity or the number of salient criteria which underlie a person's judgement. Thus, prestige and styling may be the most salient criteria a person uses when he compares various automobiles. Similarly, decay prevention and brightening of teeth may be the two criteria underlying his judgment about various brands of toothpaste.

Research Needs of Management. The above nonstatistical description of multivariate methods and their applications to marketing research problems clearly suggests that they are highly useful and relevant to marketing. Now let us also look at the potential usefulness of multivariate methods from the perspective of the major types of research needs or inputs for managerial planning. There are four types of research inputs one generally encounters in an organization. They are:

1. Diagnostic research which provides a snapshot representation of the present realities related to products or customers.
2. Prognostic research which trends or forecasts the position in which organization's products or customers are likely to be in sometime in the future.
3. Strategy research by which the organization can assess possible impact of changes in actionable programs on market behavior. This often takes the form of either field experiments or laboratory type simulations.
4. Statistical research to ensure that the quality and quantity of information to be analyzed is least biased and most satisfactorily calibrated. The statistical research essentially concerns itself with questions of sampling and nonsampling errors in data, and how to adjust for them by way of analytical strategies.

Table 1 summarizes the linkage between these four types of research needs and specific multivariate methods relevant to each of those needs. The diagnostic market research is generally exemplified by three areas of research.

The first is market segmentation based on some relevant information such as the psychographics, the demographics or the consumption patterns for which both factor analysis and cluster analysis are most appropriate techniques. The second area consists of product, brand or company typology or imagery for which also factor analysis or cluster analysis are useful techniques. The third type of diagnostic market research deals with the why aspect of customer perceptions and preferences about products for which multidimensional scaling techniques are quite relevant.

The prognostic market research is exemplified by at least two types of predictive activities. The first is the forecasting research related to company, industry or product sales either as a time series analysis or as a complex function of environmental and organizational factors. As we discussed earlier, multiple regression and canonical correlation are directly relevant for this area of prognostic research. The

TABLE 1
LINKING RESEARCH NEEDS WITH MULTIVARIATE METHODS

Research needs	Multivariate methods
A. Diagnostic research	Structural methods
1. Market segmentation	Cluster or factor analysis
2. Product or corporate typology	Factor or cluster analysis
3. Customer perceptions and preferences	Multidimensional scaling or conjoint measurement
B. Prognostic research	Functional methods
1. Sales forecasting	Multiple regression or canonical correlation
2. Market potentials	Multiple discriminant analysis
C. Strategy research	Functional methods
1. Field experiments	MANOVA or discriminant analysis
2. Laboratory simulation	MANOVA or discriminant analysis
D. Statistical research	Structural methods
1. Heterogeneity reduction	Cluster analysis or factor analysis
2. Measurement errors	Factor analysis or multidimensional scaling
3. Indexing or data consistency	Factor analysis
4. Normal distributions	Factor analysis

second area of research entails estimates of market potentials for new products as well as customer segmentation for existing products. In short, this type of research is directly related to various aspects of the product life cycle. The techniques of multiple discriminant analysis are directly relevant for this type of research.

The strategy market research often entails field experimentation or test marketing. It involves systematic manipulation of marketing mix in selected markets in order to assess their impact on market behavior such as awareness, attitudes and purchase behavior. It is obvious that multivariate analysis of variance is directly relevant here.

The final category is statistical research. There are at least four different aspects of data error and consistency. The first is the question of heterogeneity. While the present sampling theory is very useful to assist in obtaining a representative sample, there is nothing comparable to ensure a homogeneous sample. On the other hand, a heterogeneous sample has a direct adverse effect on the correlation coefficient which is often reduced to a statistical artifact of aggregating apples and oranges so to speak.

The techniques of clustering and factor analysis are, therefore, often used as intermediate stages of analysis to provide insights into the heterogeneity problem.

Another area of statistical research is concerned with the question of nonsampling measurement errors inevitable in marketing data. Often it becomes essential to eliminate this error from the data by making appropriate transformation of the data. Once again, factor analysis and multidimensional scaling become very useful intermediate procedures to remove the measurement error from the data.

The third area is concerned with the question of consistency of data. Often it is impossible to represent a complex phenomenon such as attitudes or brand loyalty by a single scale. It becomes, therefore, essential to use a variety of indicators which then must be properly indexed to produce a composite score. Once again, factor analysis becomes highly relevant for indexing purposes. Finally, often the raw data does not meet certain statistical assumptions of functional models. This is especially true with respect to the normality assumption. With the use of multivariate methods, it is possible to transform the data so that they are more normally distributed.

How to get the most out of multivariate methods

While multivariate methods have direct relevance to marketing problems as we discussed above, it is not easy to successfully implement them in the research program of the organization due to their novelty, complexity and variety. Therefore, a number of practical guidelines are described below which should be followed by the researcher if he is committed to the idea of integrating multivariate methods in his research program.

First, try not to be technique-oriented. It is not uncommon to find researchers who are comfortable with, and experienced in, a particular multivariate method such as multidimensional scaling, factor analysis, or multiple regression and try to use that technique across all research problems. They seem to be literally in search for problems which will fit the technique rather than the other way around. Often this leads to redefinition of the problem just so it meets the specifications of the technique. No single technique can solve all research problems, however, and this 'Tom Swift and his electronic machine' attitude has resulted in many misapplications of multivariate methods. While it is easy to explain this attitude as due to narrow specializations and discipline biases, it is highly hazardous to the long-term survival of multivariate methods in marketing. In fact, this technique-oriented myopic attitude of the researcher may well become the cause for the downfall of multivariate methods just as it did for operations research models several years ago [6].

Second, consider multivariate models as information inputs to managerial decisions rather than as their substitutes. Often a researcher gets carried away in building models and attempts to replace managerial judgment with the model. Unfortunately this is suicidal in view of the fact that marketing research is only a staff function whose legitimate role is to provide the necessary inputs for managerial decisions. Most managers tend to be satisficers rather than optimizers given the complexity of decisions and being continuously pressed for time. They regard research as useful input in their judgmental process but do not wish their judgment skills to be dominated by models and computers. In short, it is in the best interest of the researcher to be customer-oriented where his customers are the managers.

Third, multivariate methods or any other technique are not substitutes for researcher's skills and imagination in the proper design of the study. Statistics has nothing to do with causality and can never replace prior theory or experimental design. Unless the problem is adequately conceptualized, it is very easy in today's world of fast, efficient and inexpensive computerized calculations to evoke the GIGO principle (garbage in—gospel out)!

Fourth, half the battle in market research is proper communication of techniques and display of results. It is not at all uncommon to find a brilliant researcher totally competent in multivariate analysis whom the management or even others in the research department simply cannot understand. His communication about the beta weights, heteroscedasticity, eigenvalues, various rotations, vectors, configurations and Kruskal's Stress are Chinese and Sanskrit to the management. Consequently, the most carefully designed study with highly relevant results for managerial planning go wasted because the management simply cannot understand let alone utilize them as inputs to its decisionmaking process. It is indeed a sad state of affairs in marketing research that too little emphasis is placed on the art and science of display and communication and too much emphasis is placed on the marginal elegancies of techniques and computer programs.

Fifth, avoid making statistical inferences about the parameters of multivariate models. It is simply impossible in social sciences due to the substantial existence of nonsampling or measurement errors in the data. No sampling theory can as yet offset this nonsampling error even if one has the resources to sample the total population. Furthermore, it is not easy to apply sampling procedures in social sciences where often we don't know the population itself.

Unfortunately, too often multivariate methods have been criticized, chastized and even discarded as irrelevant tools and techniques because it is impossible to make statistical inferences. While it is true that multivariate methods require far more stringent requirements of multivariate normal distributions, it should be pointed out that distribution assumptions underlying statistical techniques even in the univariate and bivariate analysis are also impossible to meet in marketing research.

A better strategy, therefore, is not to discard the techniques as irrelevant but to put them to use for other purposes such as for making substantive inferences or as descriptive statistical techniques by which large data sets can be reduced to meaningful and concise summaries for managerial inputs. In other words, multivariate methods are more useful as data transformation, data reduction and as data display techniques than as mathematical models. This is not the fault of the techniques but the limitations of existing methods of data collection.

Sixth, guard yourself against the danger of making substantive inferences about market realities which may be an artifact solely due to the peculiarities of a particular multivariate method. Since multivariate methods are more complex statistical procedures, there are many more underlying assumptions required for the optimization (minimization or maximization) of statistical decision rules. Consequently, it is easier to inject substantive meanings in the data even if the data are essentially random relationships. This has been especially true of those multivariate methods such as cluster analysis, multidimensional scaling and conjoint measurement which possess no underlying sampling theory, and therefore, are essentially heuristics often no better than naive judgmental rules.

In order to guard against this danger, it is recommended that the same data be subjected to at least two different techniques. Often, this may be limited to two or more variations of the same basic multivariate method. The replication principle underlying this recommendation will at least bring to the researcher's attention the presence of a technique artifact in his data analysis.

Finally, exploit the complimentary relationship inherent between the structural and the functional multivariate methods. For example, it is extremely advantageous to subject the original predictor variables to a factor analysis and utilize the transformed factor scores as derived predictor variables in a multiple regression because it makes the data closer to the requirements of lack of multicollinearity and nonsampling error and the presence of normality of the distribution.

Similarly, it is best to utilize cluster analysis first to define the number of mutually exclusive groups or segments before attempting a multiple discriminant analysis. In short, this guideline urges the researcher to replace or at least substantiate a number of judgments he has to make in order to build functional multivariate methods, with a structural multivariate analysis of the data. For often the researcher's judgment is highly tenuous and sometimes patently wrong which increases the probability of building less useful multivariate models.

In conclusion, multivariate methods are highly relevant to marketing problems. However, due to lack of familiarity with them, their innate complexity and large variety, it is easy to misapply these techniques.

Several practical suggestions have been made in the paper to increase the likelihood of getting more out of the multivariate methods. Perhaps the single most important guideline to recommend is: don't be enamoured by them.

REFERENCES

1. Banks, Seymour. 'Some Correlates of Coffee and Cleanser Brand Shares', *Journal of Advertising Research*, 1 (June 1961), p. 22–28.
2. Ferber, Robert. *Handbook of Marketing Research*. New York: McGraw-Hill Book Company, 1974.
3. Gatty, Ronald. 'Multivariate Analysis for Marketing Research: An Evaluation,' *Applied Statistics*, 15 (November 1966), p. 146–58.
4. Sheth, Jagdish N. 'Multivariate Analysis in Marketing,' *Journal of Advertising Research*, 10 (February 1970), p. 29–39.
5. Sheth, Jagdish N. 'Multivariate Revolution in Marketing Research,' *Journal of Marketing*, 35 (January 1971), p. 13–19.
6. Sheth, Jagdish N. 'Some Thoughts on the Future of Marketing Models,' unpublished Faculty Working Paper No. 232, February 1975, University of Illinois.
7. Sheth, Jagdish N. *Multivariate Methods For Marketing Research*, Chicago: American Marketing Association (in press).
8. Ziff, Ruth, 'Psychographics for Market Segmentation,' *Journal of Advertising Research*, 11 (April 1971), p. 3–10.

2

Multiple Regression Analysis

CHAPTER REVIEW

This chapter describes multiple regression analysis as it is used to solve important research problems, particularly in business. Guidelines are presented for judging the appropriateness of multiple regression for various types of problems. Suggestions are provided for interpreting the results of its application, from both a managerial as well as a statistical viewpoint. Many readers will be familiar with multiple regression procedures prior to reading this chapter. It is suggested, however, that you review and familiarize yourself with the Definitions of Key Terms before reading the chapter.

Multiple regression analysis is a general statistical technique used to analyze the relationship between a single dependent variable and several independent variables. After studying the overview of regression analysis presented in this chapter, you should be able to do the following:

- [ ] Determine when regression analysis is the appropriate statistical tool to use in analyzing a problem.
- [ ] Understand how regression analysis helps us predict.
- [ ] Understand the least squares concept of prediction.
- [ ] Be aware of the important assumptions underlying regression analysis.
- [ ] Interpret the results of regression from both a statistical viewpoint and a managerial viewpoint.
- [ ] Explain the difference between stepwise and simultaneous regression.
- [ ] Use dummy variables with an understanding of their interpretation.

DEFINITIONS OF KEY TERMS

INTERCEPT. The value on the y axis (criterion variable axis) where the line defined by the regression equation $y = b_0 + b_1x_1$ crosses the axis. It is described by the constant term b_0 in the equation. The intercept may have no managerial interpretation and serve only to aid in prediction. For example, if the x_1 variable has the value zero; the y value would be predicted to be $y = b_0$. If it is not possible for x to have a value of zero this may have no meaning.

ERROR OR RESIDUAL. Seldom will our predictions be perfect. We assume that random error in prediction will occur, but we assume that this error is an estimate of the true random error in the population, not just the error in prediction for our sample. We assume that the error in the population that we are estimating is distributed with an average value of zero and a constant variance. The error in predicting our sample data is called the 'residual'.

CORRELATION COEFFICIENT (r).   Measures the strength of association between the criterion and predictor(s). The magnitude of the coefficient is not easy to interpret (see definition of Coefficient of Determination), but the sign ($+$ or $-$) indicates the direction of the relationship. The correlation coefficient varies from -1 to $+1$ with, for example, $+1$ indicating a direct perfect relationship, zero indicating no relationship, and -1 indicating a reversed relationship (as one gets larger the other gets smaller).

COEFFICIENT OF DETERMINATION (r^2).   Measures the proportion of the variation of the criterion variable about its mean which is "explained" by the predictor variable(s). The coefficient can vary between 0 and $+1$. If the regression model is properly applied and estimated, the higher the value of r^2, the greater the explanatory power of the regression equation, and therefore the better the prediction of the criterion variable.

PARTIAL CORRELATION COEFFICIENT.   Measures the strength of the relationship between the criterion variable and a single predictor variable when the effects of the other predictor variables in the model are held constant. For example, r_y, $x_2.x_1$ measures the variation in y associated with x_2 when the effect of x_1 on both y and x_2 is held constant.

HOMOSCEDASTICITY.   When the variance of the error terms (e_i) appears constant over a range of x values, the data are said to be homoscedastic and therefore satisfy the assumption of homoscedasticity. The assumption of equal variance of the population error (ε_i), where ε_i is estimated from e_i is critical to the proper application of linear regression. When the error terms have increasing or modulating variance, the data is heteroscedastic. The discussion examining the residuals (e_i) in this chapter will further illustrate this point.

COLLINEARITY.   A concept that expresses the relationship between two (collinearity) or more variables (multicollinearity). Two variables are said to exhibit complete collinearity if their correlation coefficient is one, and complete lack of collinearity if their correlation coefficient is zero.

DUMMY VARIABLES.   An independent variable used to account for the effect that different levels of a variable produce upon a dependent variable. Variables that are sometimes treated as dummy variables in regression analysis are sex, race, etc. To account for L levels of such a variety, L-1 dummy variables are needed. For example, we could represent the variable sex as two variables X_1 and X_2. When the respondent is male, $X_1 = 1$ and $X_2 = 0$. When the respondent is female, $X_1 = 0$ and $X_2 = 1$. However, when $X_1 = 1$, X_2 must be zero so we only need one dummy variable X_1 to represent the variable sex. For those students familiar with matrix algebra, b_0 of the regression equation could not be estimated because there would be redundant information. If X_1 is zero, then we know the value of X_2 has to be 1 so we don't need both values. We will always have one less dummy variable than the levels of the variables we are using.

LINEARITY. Used to express the concept that the model possesses the properties of additivity and homogeneity. In the population model $Y = \beta_0 + \beta_1 X_1 + \varepsilon$, the effect of a change of 1 in X_1 is to *add* β_1 (a constant) units to Y. The model $Y = \beta_1 X_1 {}^{\beta_1}$ would not be additive because a unit change in X_0 does not increase Y by β_1 units, rather it increases Y by $\beta_0 (X_1 + 1)^{\beta_2}$ units. (A variable amount for varying levels of X.)

ZERO SLOPE. The presence of zero slope in a regression equation indicates that the regression line is horizontal, and therefore y does not vary with x. The availability of a such a regression equation to predict values of y from values of x does not increase the predictive accuracy of the researcher. That is, if the equation describes a line with zero slope, all values of x would be associated with the same value of y.

DEGREES OF FREEDOM. Calculated from the total number of observations minus the number of parameters estimated from those data. These parameter estimates are restrictions on the data, since once made they define the population from which the data are assumed to have been drawn. For example, in estimating the random error in a two variable regression model we must have estimated two parameters, i.e., β_0 with b_0 and β_1 with b_1 since the measure of error is $\Sigma(y - \text{prediction})^2$, which is $\Sigma (y - (b_1 X)^2$. Therefore, with n values of y we have (n-2) degrees of freedom for the estimation of random error.

PARAMETER. A quantity (measure) characteristic of the population. For example, μ and σ^2 are the symbols used for the population parameters, mean (μ) and variance (σ^2). These are typically estimated from sample data where the arithmetic average of the sample is the estimate of the population average, and the variance of the sample is used to estimate the variance of the population.

REGRESSION COEFFICIENT.   The numerical value of any parameter estimate that is directly associated with the independent variables, e.g., in the model $y = b_0 + b_1 X_1$ the value b_1 is the regression coefficient for the variable X_1.

STANDARDIZATION.   The process whereby raw data is transformed into variables with a mean of zero and standard deviation equal to 1.0; the appropriate formula is $\frac{X_i - \overline{X}}{\hat{\sigma}}$. When data is transformed in this manner, the b_0 term in the regression equation assumes a value of zero. When the data have been standardized the term *beta coefficient* is often used to denote the regression coefficient. By multiplying beta by $S_{X_1}^2 / S_y^2$ (i.e., the beta must be multiplied by the ratio of the variance of X_1 to the variance of y), we can obtain the regression coefficient for the raw data (standardized).

PARTIAL F VALUES.   When a variable (say X_a) is added to a regression equation after many other variables have already been entered into the equation its contribution may be very small because it (X_a) is highly correlated with the variables already in the equation. The partial F test is simply an F test for the additional contribution of a variable above the contributions of those variables already in the equation. A partial F may be calculated for all the variables by simply pretending that each, in turn, is the last to enter the equation. This gives you the additional contribution of each of the variables above all others in the equation.

2

WHAT IS MULTIPLE REGRESSION ANALYSIS?

Multiple regression analysis is a statistical technique which can be used to analyze the relationship between a single dependent (criterion) variable and several independent (predictor) variables. The objective of multiple regression analysis is to use the several independent variables whose values are known to predict the single dependent value the researcher wishes to know. As was noted in the first chapter classification of multivariate statistical tools, multiple regression analysis is a dependence technique. Thus, to use multiple regression you must be able to divide the variables into a single dependent variable and several independent variables. It also should be noted that regression analysis is the statistical tool which should be used when both the dependent and the independent variables are metric. But it is possible under certain circumstances to use dummy-coded independent variables in the analysis. To summarize, this means that to use multiple regression analysis, the data must be metric and prior to derivation of the regression equation the researcher must decide which variable is to be the dependent variable and which are to be the independent variables.

WHAT DO WE DO WITH REGRESSION ANALYSIS?

As noted earlier, multiple regression analysis is a general statistical technique which can be used to examine the relationship between a single dependent variable and a set of independent variables. The following represent four different purposes that multiple regression analysis can be used to achieve:

1. Determine the appropriateness of using the regression procedure with the problem. For example, is the regression approach appropriate for attempting to predict company sales from the expenditures for advertising? The results of applying regression analysis may be interpreted in such a way as to suggest if the application was appropriate.

2. Examine the statistical significance of our attempted prediction. If we used a sample of patrons in a restaurant and attempted to predict their monthly expenditures on dining out from information on their family income, their family size, and the age of the head of household, is our prediction any better than one might expect by chance?

35

3. Examine the strength of association between the single dependent variable and the one or more independent variables. When collinearity among the independent variables is minimal (or has been removed by factor analysis), then we can identify the extent to which each of the independent variables is related to the dependent variable. For example, we can determine which variable is more important in predicting the number of ounces of hand lotion used in a household—the number of children in the household, the age of the female head of household, or the family's income level?

4. Predict the values of one variable from the values of others. Can we predict the number of cases of detergent we will sell from knowledge of the number of competitive brands in each store carrying our detergent, and the median family income in each store's trade area?

In all the previous examples one variable was given a rather special status—it became the variable we wished to predict (the dependent variable). The variables we select in our attempt to predict the criterion variable are called independent, or predictor variables. In using regression analysis a decision has to be made regarding the number of predictor variables to include in the equation. In making this decision we would assume that each additional predictor variable would give us more information and therefore a better prediction about the criterion variable. Otherwise it would not be included in the analysis.

HOW DOES REGRESSION ANALYSIS
HELP US TO PREDICT?

The objective of regression analysis is to help us to predict a single dependent variable from the knowledge of one or more independent variables. When the problem involves a single dependent variable which is predicted by a single independent variable, the statistical technique is referred to as simple regression. When the problem involves a single dependent variable predicted by two or more independent variables, it is referred to as multiple regression analysis. The following section is divided into three parts to help you understand how regression helps us to predict. The three topics covered are: (1) prediction using a single measure—the average, (2) prediction using two measures—simple regression, and (3) prediction using several measures—multiple regression.

Prediction Using a Single Measure—the Average

Let's start with a simple example. Assume we surveyed eight families and asked how many credit cards were held by all family members. The data is shown in columns one and two of Table 2.1. If we were asked to predict how many credit cards a family holds using only this data, we could simply use the arithmetic average of seven credit cards as an acceptable predictor. Our prediction typically would be stated as "the average number of credit cards held by a family is seven."

One question left unanswered is: "How accurate is this prediction?" The customary way to assess the adequacy of using the average as a predictor is to examine the errors that are made when it is used as the predictor. For example, if we predict that family number one has seven credit cards we overestimate by three. Thus the error is plus three. If this procedure were followed for each of the families, in some instances our estimate would be too high, in others too low, and for some it would give the correct number of cards held. By simply adding the errors we might expect to obtain a measure of the prediction accuracy. However, we would not—the errors would always sum to zero. Therefore, we would not have a measure of the adequacy of our prediction. To overcome this problem we can square each error and then add them together to obtain the sum of squared errors. The

TABLE 2.1
HATCO SURVEY RESULTS FOR AVERAGE
NUMBER OF CREDIT CARDS

Family Number	Actual Number of Credit Cards	Average Number of Credit Cards[1]	Error[2]	Errors Squared
1	4	7	+3	9
2	6	7	+1	1
3	6	7	+1	1
4	7	7	0	0
5	8	7	−1	1
6	7	7	0	0
7	8	7	−1	1
8	10	7	−3	9
	56		0	22

[1]Average Number of Credit Cards $= 56 \div 8 = 7$
[2]Error refers to difference between actual number of credit cards held by a family and the estimated number of cards held (7) using the arithmetic average as a predictor.

result, referred to as the "sum of squared errors," provides a good measure of the prediction accuracy of the arithmetic average. We wish to obtain the smallest possible sum of squared errors, as this would mean that our predictions would be more accurate. For a single set of observations, no approach (including other more sophisticated statistical techniques) will produce a smaller sum of squared errors than will the arithmetic average. Therefore, for a single set of observations the average is the best predictor of the number of credit cards held by families. (Note the sum of squared errors for our example problem is 22).

Prediction Using Two Measures—Simple Regression

As researchers and businesspeople, we are always interested in improving our predictions. In the preceding section we learned that with a single set of measures the average is the best predictor. But in our example survey we also collected information on other measures. Let's determine if knowledge of another measure—the number of people in each of the families—will help our predictions. This procedure involves two measures and is referred to as simple regression.

Simple regression is another procedure for describing data (just as the average describes data) and it uses the same rule—minimizing the sum of the squared errors of prediction. We know that (without using family size) we can describe the number of credit cards held as "7". Another way to write the prediction is:

predicted number of credit cards held =
average number of credit cards held

or $\hat{Y} = \bar{y}$.

Using our additional family size information, we could try to improve our predictions. We assume that if family size is related to the number of credit cards held by the family, by trying to find what difference in number of credit cards is associated with difference in family size (See Table 2.2) we can improve our prediction. We could describe the relationships as

number of credit
cards held = change in number of credit   X   (family size)
cards held associated with
unit change in family size

or $\hat{Y} = b_1 X_1$.

For example, if we find that for each additional member in a family the number of credit cards increases (on the average) by two credit cards, we would predict that families of size four would have eight credit cards, and families of size five would have ten credit cards. Thus the number of credit cards $= 2$ X (family size). However, we often find that the prediction is improved by adding a constant value to our prediction because the following relationship may be found:

Family Size	Average Number of Credit Cards
1	4
2	6
3	8
4	10
5	12

It can be observed that "number of credit cards $= 2$ X (family size)" is wrong by 2 credit cards in every case, e.g.:

Family Size	Average Number of Credit Cards	Predicted Number of Credit Cards	Error
1	4	2	-2
2	6	4	-2
3	8	6	-2
4	10	8	-2
5	12	10	-2

Therefore, changing our description to:

number of credit cards $= 2 + 2$ X (family size)

gives us perfect predictions in all cases.

We will take this approach with our sample of eight families and see how well the description fits our data. The procedure followed will be:

predicted number of credit
cards = constant + change in number   X   (family size)
of credit cards
with differing
family size

or $\hat{Y} = b_0 + b_1 X_1.$

TABLE 2.2
HATCO SURVEY RESULTS RELATING NUMBER
OF CREDIT CARDS TO FAMILY SIZE

Family Number	Number of Credit Cards[1] (Y_i)	Family, Size[2] (X_1)	Prediction	Error Squared
1	4	2	4.81	.66
2	6	2	4.81	1.42
3	6	4	6.76	.58
4	7	4	6.76	.06
5	8	5	7.73	.07
6	7	5	7.73	.53
7	8	6	8.7	.49
8	10	6	8.7	1.69
				5.50

[1] Average Number of credit cards $= \Sigma \; ^x/n \; 56/8 = 7$
[2] Average Family Size $= \Sigma \; ^x/n = 34/8 = 4.25$

If the constant term (b_0) does not help us predict, then the process of minimizing the sum of squared errors will give an estimate of the constant term to be zero. The terms b_0 and b_1 are called regression coefficients.

Using either trial-and-error or a faster mathematical procedure,[1] we find the values of b_0 and b_1 such that the sum of the squared errors of prediction are minimized. For this example the appropriate values are:

$$\hat{Y} = 2.87 + .97 \text{ (family size)}.$$

Since we have used the same criterion (least squares) we can see if our knowledge of family size has helped us predict credit card holdings. The sum of squared errors using the average was 22, and the sum of squared errors using our new procedure is 5.50 (see Table 2.2, column 5). Using the least squares criterion we can say that our new approach—simple regression—is better than using the average. This description indicates that for each additional family member the credit card holdings are on the average higher by .97 credit cards. The constant term 2.87 is an artificial term useful for helping us pre-

---

[1] The mechanics of deriving the regression coefficients such that the sum of the squared errors is minimized is left to other more technically oriented texts dealing with regression. See [1, 2, 3, and 4].

dict, but has no interpretation in credit card holding because we do *not* assume that if family size was zero, credit card holdings of 2.87 would exist.

Major Assumptions We Have Made

We have shown how improvements in prediction are possible, but in doing so we had to make several assumptions about the relationship between the variable to be predicted and the variables we want to use for predicting. In the following sections we discuss these assumptions, which include: the assumption of a statistical relationship, the assumption of equal variance and the assumption of uncorrelated errors.

Statistical Relationship. Since we are dealing with sample data representing human behavior, we are assuming that our description of credit card holdings is statistical, not functional. For example:

$$\text{Total Cost} = \text{Variable Cost} + \text{Fixed Cost}.$$

If variable cost is $2.00 per unit, fixed cost is $500.00, and we produce 100 units, we assume that total cost will be exactly $700.00 and any deviation from $700.00 is caused by our inability to measure cost *because the relationship between the costs is fixed*. This is called a functional relationship.

In our credit card example, we found two families with two members, two with four, etc., who had different numbers of credit cards. More than one value of the criterion variable will usually be observed for any value of a predictor variable. The criterion variable is assumed to be a random variable (see Definitions) and for a given predictor we can only hope to *estimate the average value* of the criterion variable associated with the predictor. In our example, the two families with four members held an average number of 6.5 credit cards and our prediction was 6.76. Our prediction is not as accurate as we would like, but it is better than just using the average of 7 credit cards. The error is assumed to be the result of random behavior among credit card holders.

In summary, a functional relationship calculates an exact value while a statistical relationship estimates an 'average' value. We will be concerned throughout this text with statistical relationship. Both of these relationships are displayed in Figure 2.1.

Equal Variance of the Criterion Variables. In most situations, we will have many different values for the criterion variable at each

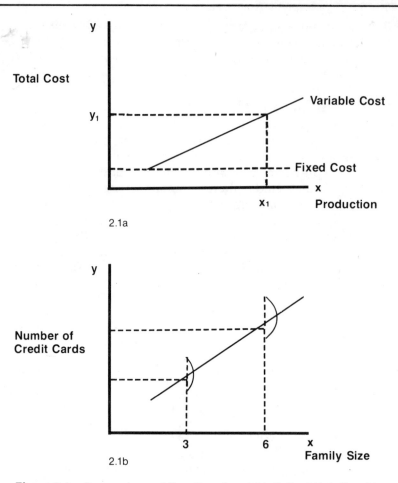

Figure 2.1—Comparison of Functional and Statistical Relationships

value for the predictor variable just as we had two different numbers of credit cards held for our two families of four members. Since our prediction is based upon actual values from our predictions, we assume that at each level of the predictor variable the values of the criterion variable all have the same variance. This is illustrated in Figure 2.1b by showing the distributions of the criterion variable to be identical at the two levels of the predictor variable. The illustration indicates that families of size 3 and size 6 have many different numbers of credit cards among them, but the numbers of credit cards tends to vary from the average held by families of size three by the same amount that the number held by families of size six vary from their average number held. If this is not true, then our predictions are going to be better at some levels of the criterion variable than at others. However, since our rule for determining how good our predictions are is based on the equated deviations over *all* levels of the predictor variable, we could be misled into thinking our predictions have approximately equal descriptiveness at all levels.

Errors are Uncorrelated. Our predictions were not perfect in our credit card example, and we will rarely find a situation where they are perfect. However, we do want to find that any errors we make in prediction are uncorrelated with each other. If we found a pattern that suggests every other error is positive while the alternative error terms are negative, we would know that some unexplained systematic relationship exists in the criterion variable. If such a situation exists, when we use our predictions we cannot have confidence that our prediction errors are independent of the levels at which we are trying to predict. In our credit card example, we would like to find just as strong a belief that any error is randomly distributed when predicting for a family of size 2 as for a family of size 6.

Prediction Using Several Measures—
Multiple Regression Analysis

We previously demonstrated how simple regression helped improve our prediction of credit card holdings. By using data on family size we could more accurately predict the number of credit cards a family would have than we could by simply using the arithmetic average. This raises the question of whether we could improve our prediction even further by using additional data obtained from the families. Would our prediction of the number of credit cards be improved if we used data not only on family size but also on family income?

In an effort to further improve our prediction of credit card holdings, let's use additional data obtained from our eight families. The variable we shall add is family income (See Table 2.3).

We simply expand our simple regression model to:

| Number of Credit Cards. | = | Constant number of credit indepen- dent of family size and in- come. | + | (Change in credit card holdings associated with unit change in family size.) | × | (Family Size) | + | (Change in credit card holdings associated with unit change in family size.) | × | (Family income) |

or $Y = .482 + .63X_1 + .216X_2$.

TABLE 2.3

HATCO SURVEY RESULTS RELATING NUMBER OF CREDIT CARDS
TO FAMILY SIZE AND FAMILY INCOME

Family Number	Number of Credit Cards (Y)	Family Size (X_1)	Family Income (X_2)	Prediction	Error Squared
1	4	2	14	4.76	.59
2	6	2	16	5.20	.64
3	6	4	14	6.03	.00
4	7	4	17	6.68	.10
5	8	5	18	7.53	.22
6	7	5	21	8.18	1.38
7	8	6	17	7.95	.00
8	10	6	25	9.67	.11
					3.05

We can again find our error by predicting Y and subtracting our prediction from the actual value as in columns 5 and 6 of Table 2.3. The total sum of squared error is 3.05 for our prediction using both family size and family income while it is 5.50 (Table 2.2) using family size and 22 (Table 2.1) using the arithmetic average. We assume at this point that some improvement in prediction has been found.

What New Assumptions Have We Made?   We have added one more predictive variable to better predict the number of credit cards held by families. When doing this we must keep in mind all of the previously cited assumptions, and must concern ourselves with any possible interaction and/or correlation among our predictor variables because we assume they do not interact and are uncorrelated. The rationale for considering independence, interaction, and/or cor-

TABLE 2.4

NUMERICAL ILLUSTRATION OF INDEPENDENT,
INTERACTING AND CORRELATED PREDICTORS

2.4a Independent			2.4b Interacting			2.4c Correlated		
Credit Cards	Family Size	Family Income	Credit Cards	Family Size	Family Income	Credit Cards	Family Size	Family Income
3	2	10	2	2	10	2	2	10
4	4	10	8	4	10	4	4	20
5	2	20	6	2	20	6	2	10
6	4	20	8	4	20	8	4	20

relation among the predictor variables is discussed in the following paragraphs.

Independent. The ideal situation would be to find data such as that shown in Table 2.4a. In attempting to predict credit card holdings we would find that the following equation describes the relationship:

$$\hat{Y} = .5X_1 + .2X_2.$$

where

X_1 = Family Size

X_2 = Family Income.

We can conclude that by holding family income constant, credit card holdings change on the average by .5 for each additional member of the family. Conversely, by holding family size constant, the number of credit cards held increases on the average by .2 for a $1 increase in family income. Figure 2.2a illustrates why we can confidently make these statements. There is no correlation or interaction between the two predictor variables family size and family income. For each level of family size a change in family income of $1.00 produces a constant change in average credit card holdings. For each level of family income, a change in family size of two members produces a constant change in average credit card holdings. We can now examine the effects on our interpretation to see if there is interaction or correlation between our two predictor variables.

Interaction. The data shown in Table 2.4b reveals interaction between family size and family income, illustrated by the Figure 2.2b graph. It becomes evident that at larger family sizes, level of income has no effect on number of credit cards held. For a Family Size Two, a change in income from $10 to $20 is associated with a four-unit change in credit cards. This is not true for families of Size Four where there is no difference in the number of credit cards held. If we were to use multiple regression to predict credit card holdings we would find the following predictive equation:

$$Y = -3 + 2X_1 + .2X_2.$$

This is the best least squares equation available. It cannot be interpreted in the same way we interpreted the equation when we had no interaction. In this interactive situation, the coefficient for the effect of family size (2) reflects an average effect over both levels of family income and does not represent a constant effect. Similarly, the coefficient for family income (.2) represents an average effect of family income over the two levels of family size and again, not a constant effect.

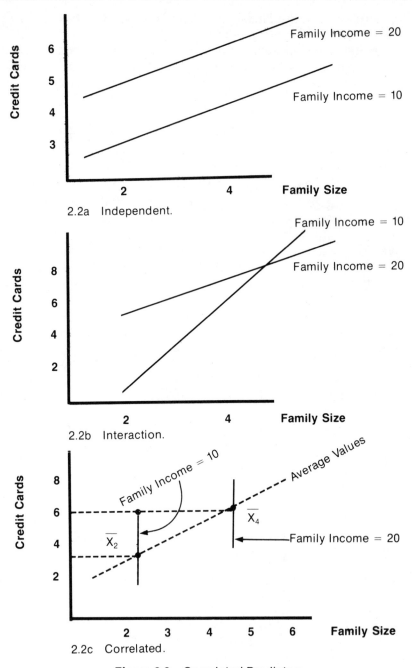

2.2a   Independent.

2.2b   Interaction.

2.2c   Correlated.

Figure 2.2—Correlated Predictors

Correlation. The data shown in Table 2.4c reflects perfect correlation between the two predictor variables, family size and family income. The simple correlation between these two variables is 1.0, as indicated in the following correlation matrix:

Correlation Matrix

	Number of Credit Cards	Family Size	Family Income
Number of Credit Cards	1.00	.45	.45
Family Size	.60	1.00	1.00
Family Income	.45	1.00	1.00

Since this correlation is perfect, we can suspect that no additional information will be gained by using both variables. You can see in Figure 2.2c that the change in credit cards held associated with a change in family size from 2 to 4 is exactly the same as the average change in credit cards held associated with a change in income from 10 to 20. In a situation such as that previously illustrated, only the predictor variable with the highest simple correlation (family size) would be used to estimate the number of credit cards held by a family. Other variables that are highly correlated with the single best predictor would be discarded. The *rule-of thumb* often utilized to determine the cut-off point for intercorrelation among predictor variables is that no predictor should be included which is more closely related to the best predictor than it is with the dependent variable.

WHERE CAN WE GO FROM SIMPLE PREDICTION?

In each illustration we have been concerned with predicting the number of credit cards held by families and have seen that the arithmetic average, family size simple regression, or family size and family income multiple regression all provide predictions. Further, we have been able to assess the accuracy of our predictions by examining the sum of the squared errors of prediction. This is an oversimplification, but will serve to acquaint you with the essential meaning of regression analysis. Now, we must examine our predictive ability from a statistical viewpoint (as an estimator of population characteristics.).

In our example of multiple regression, we used a sample of eight families and attempted to predict the number of credit cards they held using two other measures (family size and family income). Rarely would we want to know only about these eight families. Instead, in most cases we would take a sample to develop a predictive model for *all* families having credit cards. From this viewpoint our predictions about the eight families are only a basis of obtaining the predictive model $(Y = b_0 + b_1X_1 + b_2X_2)$. We now become concerned with how well we think this model predicts for all families having credit cards. That is, how well will our model predict for families not included in the sample we used to develop the predictive equation? We believe that a relationship between family size, family income, and number of credit cards exists in the *population of all* families having credit cards, and our sample of eight families allows us to infer what the relationship is in the population.

Let's review four different purposes that multiple regression analysis can achieve:

(1) Determining the appropriateness of our predictive model.
(2) Examining the statistical significance of our model.
(3) Predicting with the model.
(4) Examining the strength of association between the variables.

All the above topics will be approached from a viewpoint of understanding the characteristics of the population based on the sample data rather than simply examining the sample data. For many illustrations we will use the simple model (predicting credit cards from family size) because it is easier to visualize than the more complicated model with two predictor variables (family size and family income).

Determining the Appropriateness of our Predictive Model

After calculating the regression coefficients and looking at the predictions made with our equation, we are faced with a dilemma. Have we met the assumptions of the model? Are there errors in prediction that suggest we look for new predictive variables? When the model is completed we have to examine its appropriateness to answer these questions. This can be accomplished through an examination of the errors in prediction—called the "residuals."

We should note that residuals (the differences between the observed values and the values our model predicts) are an artifact of the particular predictive model we are using and not equivalent to the random error in the population. These residuals should reflect the properties of the population random error if the model is appropriate. Analysis of residuals may therefore be used to examine the appropriateness of the predictive model in terms of:

1. the linearity of the phenomenon measured,
2. the constant variance of the error terms,
3. the independence of the error terms,
4. the normality of the error term distribution, and
5. the addition of other variables.

Linearity of the Phenomenon. We can examine the linearity of the phenomenon in two ways: partitioning the error, and plotting the residuals. Plotting the residuals against the X values (family size) often reveals the shortcomings of inappropriately applying a linear model. The plot should show the residuals falling randomly, with relatively equal dispersion about zero, and no strong tendencies to be either greater or less than zero.

Partitioning the error consists of examining the residuals to estimate how much of the variation is due to lack of fit of the model, and how much of the variation is due to random error. That is, how well do we predict the average value of Y for each value of X? To do this we must have repeated observations for at least two of the X values (family size) and hopefully for all of them. For every family size (x) for which we have multiple observations, we use the average number of credit cards rather than the actual data. In our study we had two families of each size, so we would use the data shown in Table 2.5. We would use our predictive model ($Y = b_0 + b_1X_1$) and calculate the sum of squared errors for the averages. Since we originally stated that the model will predict the *average* number of credit cards held, this analysis, using only the means, directly examines this prediction. From our data (Table 2.5) the squared errors of prediction are .486. Since our sum of squared errors using the original data was 5.50 we can see that we are predicting the average number of credit cards for each family size rather well. The error in predicting the average values for each family size is called a "lack of fit" error while the variation of

TABLE 2.5
HATCO SURVEY RESULTS FOR AVERAGE NUMBER
OF CREDIT CARDS AND FAMILY SIZE

Family	Average Number of Credit Cards per Family Size	Prediction	Family Size	(Prediction Error Squared)
1	5	4.8	2	.036
2	5	4.8	2	.036
3	6.5	6.7	4	.063
4	6.5	6.7	4	.063
5	7.5	7.7	5	.048
6	7.5	7.7	5	.048
7	9	8.7	6	.096
8	9	8.7	6	.096
				.486

the individual observations about these averages is called "pure" error. Figure 2.3 illustrates lack of fit and pure error. We cannot possibly hope to predict each observation (pure error) but we do hope to predict all the averages (lack of fit).

Intuitively one sees that if the pure error is significantly less than the lack of fit error, the appropriateness of the predictive model is questionable. We can test the appropriateness of our model with the F test which simply compares the lack of fit to the pure error. For our data the F value $(5.00 \div 4\text{d.f.}/ .50 \div 2\text{d.f.})^2$ is .196. An F value approaching 1 indicates that a linear model is appropriate while large values of F suggest that the phenomenon is not linear. If the model fits the data perfectly, then lack of fit error equals zero and the only error is the pure error—a rare occurrence.

Constant Variance of the Error Term. A plot of residuals against the predicted criterion variable indicates that the variance is not constant if it displays anything other than a random pattern. Figure 2.4a displays a hypothetical random distribution of residuals. Thus, the variance of the error term is constant and a linear regression model is appropriate.

Independence of Error Terms. Plotting the residuals against time, even if time is not a value under consideration for the model, will reveal whether the sequencing of the measurements has affected the outcome of the experiment. Again, the pattern should appear random.

Figure 2.4b displays a residual plot which exhibits an association between the residuals and the sequencing of the measurements. That is, early measurements are negative, but later ones increase and become positive.

Normality of Error Term Distribution. Three procedures can be used to test the normality of error term distribution. The simplest method of testing is the construction of histograms of the residuals to visually check if the residuals appear to have a normal distribution. Alternatively, one might determine the percentages of the residuals falling within $\pm$ one standard error or $\pm$ two standard errors. A third

---

[2] Since we have based the analysis on the four average values for number of credit cards held, we have estimated four parameters and therefore have only the 8 observations minus the four averages $(8-4=4)$ degrees of freedom for the numerator of the F test. The denominator has 2 degrees of freedom because we have used the four averages to estimate the equation parameters of β_0 and β_1. (i.e. the number of average values minus the number of parameters is $4-2=2$ degrees of freedom.) In practice one would not attempt an analysis with this small a data set. An often used rule of thumb is to have 10 times more observations than variables in the predictive model.

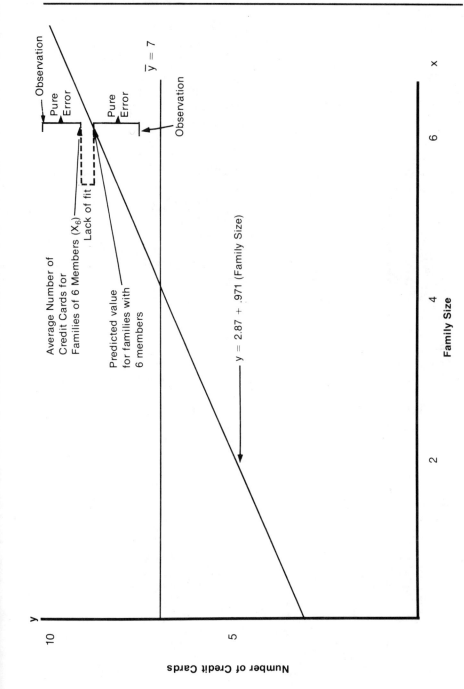

Figure 2.3   Illustration of pure error and lack of fit error

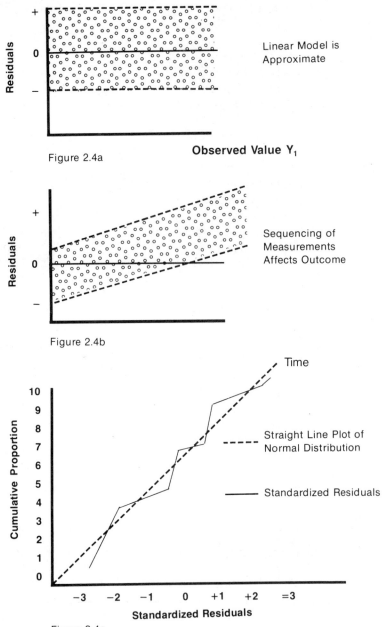

Figure 2.4a

Figure 2.4b

Figure 2.4c

Adapted from John Neter and William Wasserman, *Applied Linear Statistical Models*, Richard D. Irwin, Inc., 1974.

Figure 2.4   Residual plots

procedure would be to plot the cumulative standardized residuals on normal probability paper as in Figure 2.4c. Any departure from normality may be seen by comparing the cumulative standardized residuals with the straight line representing the perfect normal distribution. To do so, the analyst would examine the plots to determine if they are reasonably comparable (without extensive deviations).

Addition of Other Variables. After examining our predictive model we might consider including another variable such as family income as illustrated earlier. One way to decide whether or not to add the new variable is to take only errors (residuals) from our predictions using family size and try to predict these errors using our new variable. If the new variable(s) explain a significant portion of the residual variation, it should be examined carefully for inclusion.

In summary, residuals give the analyst a quick avenue for determining how well the model approximates the assumptions upon which it is based and what types of errors are being encountered. It also provides a vehicle for examining other variables as candidates for inclusion.

Examining the Statistical Significance of Our Model

If we were to take repeated samples of eight families and ask them how many family members and credit cards they have, we would seldom get exactly the same values for $Y = b_0 + b_1 X_1$ from all samples. We would expect chance variation to cause differences among many samples. Usually, we take only one sample and base our predictive model on that sample. We can test certain hypotheses concerning our predictive model to ensure it does represent the population of all families having credit cards rather than just our one sample of eight people. These tests may take two basic forms: a test of coefficients and a test of the variation explained (coefficient of determination).

Tests of Coefficients. In the sample model, we said the number of credit cards equaled 2.87 + .971 (family size). We would test two hypotheses:

Hypothesis 1:

The intercept (constant term) value of 2.87 arose by sampling error and the "real" constant term appropriate to the population is zero. With this hypothesis we would simply be testing whether the constant term should be considered appropriate for our predictive model. If it is found not to differ significantly from zero, we would assume the constant term should not be used for predictive purposes. The appropriate test is the t test and is commonly available on computerized regression analysis programs.

Hypothesis 2:

The coefficient .971 indicates that an increase of one unit in family size increases the average number of credit cards held by .971 cards. We can test if this coefficient differs significantly from zero. If this coefficient (.971) could have occurred because of sampling error, we would conclude that family size has no impact on number of credit cards held. You should note this is not a test of any exact value of the coefficient, but rather a test of whether it should be used at all. Again, the appropriate test is the t test.

Coefficient of Determination.   A quick way to obtain an approximation of how well the line we have fitted describes family holdings of credit cards is to examine the amount of variation our predictive model explains. For example, we could take the average number of credit cards held by our sampled families and use it as our best estimate of the number held by any family. We know that this is not an extremely accurate estimate, and we examined its accuracy by calculating the squared sum of errors in prediction (Sum of squares = 22). This is a measure of how well the average explains the purchases observed. Now that we have fitted a regression model using family size, does it explain the variation better than the average? We know it is somewhat better because the sum of squared errors is now 5.50. We can look at how well our model does by examining this improvement:

Sum of squared errors in prediction
using the average = 22.0

Sum of squared errors in prediction
using family size =   5.5

_____

Sum of squared errors "explained"
by family size = 16.5

Therefore, we explained 16.5 squared errors by changing from the average to a regression model using family size. This is an improvement of 75 percent (16.5 ÷ 22 = .75) over use of the average.

The ratio of

Sum of Squared Errors 'Explained' by Regression

_____

Sum of Squared Errors About the Average

is called the *coefficient of determination*. The symbol r^2 is used to represent the coefficient of determination. If the regression model using family size perfectly predicted all families holdings of credit cards, this ratio would equal 1. If using family size gave no better predictions than using the average, then the ratio (r^2) would be very close to zero.

We often use the coefficient of correlation (r) to assess the relationship between Y and X. Other than having a sign ($+r, -r$) to denote the slope of the regression line, it offers little in clarifying the nature of the association. For example, if $r^2 = .75$, we know that 75 percent of the variation in Y is explained by introducing the variable X. The corresponding value of $r = +.86$ offers the sign ($+$) as additional information but may mislead some analysts to believe a stronger relationship exists.

To test the hypothesis that the amount of variation explained by the regression model above that explained by using the average for prediction did not occur by chance (i.e., that r^2 is greater than zero), the F ratio is used. The test statistic F is the ratio of:

Sum of Squared Error 'Explained' by Regression $\div$ Degrees of Freedom

---

Sum of Squared Error About the Average $\div$ Degrees of Freedom

Two important features of this ratio should be noted:
1. Each sum of squares divided by its appropriate degrees of freedom is simply the variance of the prediction errors.
2. Intuitively, one sees that if the ratio of the 'explained' variance to the variance about the mean is high, the use of family size must be of "significant value" in explaining the number of credit cards held by families.

For our example, the F ratio is $(16.5 \div 1/5.50 \div 6) = 18.1$. When compared to the Tabled F statistic for 1 and 6 degrees of freedom of 5.99 (which would occur with probability of .05), it leads us to reject the hypothesis that the reduction in error we obtained by using family size to predict credit card holdings was a chance occurrence. This means that finding a sample showing we can explain 18 times as much variation using family size, as when using the average, is not very likely to happen by chance (less than 5 percent of the time) when family size is *not* related to number of credit cards in the population.

Predicting With the Model

After satisfying ourselves that using family size to predict number of credit cards offers predictions significantly better than those afforded by the average, we examine the actual predictions. This is easily done by plugging in the appropriate values, as in the case of a family size of six for our example:

$$\hat{Y}_i = b_0 + b_1X_i$$
$$\hat{Y}_i + 2.87 + .971(6) = 8.7$$

Since 8.7 is the estimated mean value for the number of credit cards when the family size is six, we must look at the potential error in this estimate to understand its value. If we were to take repeated samples at $X=6$ we know we would observe differing values of Y (our original data supports this). When predicting as was just described, you are predicting an *average* value of Y that might occur for a given value of X. In our problem we could predict that for families of size six, the *average* number of credit cards would be 8.7 cards. This *does not* mean that we would expect every family of six people to have 8.7 credit cards. It just means that our best estimate of the average number held by families of six members is 8.7 cards. We can place confidence limits about the estimated value of Y just as we can place confidence limits on other estimates from samples. These confidence limits would give us a range in which we expect the average for our sample values to fall relative to the true population characteristics.

Since we estimated our model from sample values, the estimates b_0 and b_1 are estimated from values of both Y and X. If the values of X (in our problem: family size) differ greatly from the average family size ($\overline{X}=4.25$) from sample to sample, the estimates of b_0 and b_1 can vary greatly. As we make estimates of Y for families of sizes further from $\overline{X}$, the confidence range on the predictions gets wider. As a rule of thumb, if you wish to predict from a regression model, select values of X such that your area of prediction is near the average ($\overline{X}$) value of X. This means that if you are especially interested in predicting for medium sized families of three, four, or five members, try to include families both larger and smaller in the sample. Additionally, one assumes that the predictions are valid only if:

a. The conditions and relationships measured at the time the sample of families is taken do not materially change. For example, if most companies suddenly start charging a monthly

fee for credit cards, your predictions could be very poor because families may change their credit card holdings.
b. The model is not to be used for estimation beyond the range
of the X values (family size) found in the sample. For instance,
if the largest family in your sample had six members, it might
be unwise to predict the number of credit cards held by families of ten members.

In the former case, one is assuming that the model continues to fit
the modeled relations over time. In the latter case, one assumes (dan-
gerously) that values of X beyond those measured will have the same
relation with Y as those X values included in the regression model.
In both situations, you have few ways to legitimately validate these
assumptions other than re-estimating the models with new data.

Examining Strength of Association Among Our Variables

Using the Regression Coefficients. We would like to know which
variable—family size or family income—helps us the most to predict
the number of credit cards held by a family. Unfortunately, the regression coefficients (b_0, b_1, and b_2) do not really give us this information.
To illustrate why, we can use a rather obvious case. Suppose we wanted
to predict teenagers' monthly expenditures on records (Y) using two
predictor variables; X_1 is parent's income in thousands of dollars and
X_2 is the teenager's monthly allowance measured in dollars. We found
the following model by a least squares procedure:

$$Y = -.01 + 1X_1 + .001X_2$$

You might assume that X_1 is more important because its coefficient is
1000 times larger than the coefficient for X_2. This is not true however.
A $10 increase in parents income produces a 1 X $10 ÷ $1000
change in average record purchases (since the X_1 value is measured in
thousands of dollars). This change is .01 in the average number of
records. A change of $10 in monthly allowance for teenager produces
a (.001) ($10) change in average record expenditures or a .01 change
in average number of records (as teenagers allowance was measured
in dollars).

A $10 change in parent's income produced the *same* effect as a
$10 change in teenager's allowances. Both variables are equally im-
portant, but the regression coefficients do not directly reveal this. We
can resolve this problem by using a modified regression coefficient
called the beta coefficient.

Beta Coefficient. If each of our predictor variables had been standardized (see chapter definitions) before we estimated the regression equation, we would have found different regression coefficients. The coefficients resulting from standardized data are called *beta coefficients*. Their value is that we no longer have the problem of different units of measure (as illustrated previously), and the beta coefficient reflects the impact on the criterion variable of a change of one standard deviation in either variable. Now we have a common unit of measure, and the coefficients tell us which variable is most influential.

Three cautions must be given when using beta coefficients. First, beta coefficients should be used as a guide to the relative importance of individual independent variables only when collinearity is minimal. Second, the beta values can only be interpreted in the context of the other variables in the equation. For example, a beta value for family size only reflects its importance relative to family income, not in any absolute sense. If another predictor variable was added to the equation, the beta coefficient for family size will probably change, because there will likely be some relationship between family size and the new predictor variable.

The third caution is that the levels (e.g. families of size 5, 6 and 7) affect the beta value. Had we found families of size 8, 9 and 10, the value of beta would likely change. In summary, use beta only as a guide to the relative importance of the predictor variables included in your equation, and only over the range for which you actually have sample data.

GENERAL APPROACHES TO REGRESSION ANALYSIS

There are many approaches you may use when attempting to determine the "best" predictive model using regression analysis. The two most common are backward elimination and stepwise forward estimation. These procedures and some general guidelines for using them will be discussed next.

Backward Elimination

This procedure is largely a trial-and-error procedure for finding the "best" regression estimates. It involves computing a regression equation with all the variables, and then going back and deleting those independent variables which do not contribute significantly to the equation. The specific steps are:

1. Compute a single regression equation using all of the predictor variables which interest you.
2. Calculate a partial F test (see definitions) for each variable as if it were to be used *after the variance for all other variables* is removed.
3. Eliminate that (or those) predictor variable with a partial F value you judge indicative that the variable under consideration is not making a significant contribution.
4. After eliminating variables, re-estimate the regression model using only the remaining predictor variables.
5. Return to step 2 and continue the process until you identify all variables that make a significant contribution.

This procedure for examining the regression model allows you to examine all variables of interest and see what their contributions are. It is time-consuming, but with adequate computer facilities it is a satisfactory process for many researchers.

Stepwise Forward Estimation

This procedure also allows you to examine the contribution of each predictor variable to the regression model, but rather than deleting variables as in the backward elimination procedure, each variable is considered for inclusion prior to developing the equation. The specific steps are:

1. Start with the simple regression model in which *only* the one predictor most highly correlated with the criterion variable is used. The equation would be: $Y = b_0 + b_1 X_1$.
2. Examine the partial correlation coefficient (see definitions) to find an additional predictor variable that explains both a significant portion and the greatest portion of the error remaining from the first regression equation.
3. Recompute the regression equation using the two predictor variables, and examine the partial F value for the original variable in the model to see if it still makes a significant contribution, given the presence of the new predictor variable. If the original variable still makes a significant contribution you would have the equation:

$$Y = b_0 + b_1 X_1 + b_2 X_2.$$

4. Continue this procedure by examining all predictors not in the model to determine if one should be included in the

model. If a new predictor is included, examine all predictors in the model previously to judge if they should be kept.

In the following section an application of stepwise regression is shown to further illustrate these concepts.

ILLUSTRATION OF A REGRESSION ANALYSIS

In Chapter One you were introduced to a problem in which HATCO had obtained the following measures:

Variable Description
X_1 Self-esteem
X_2 Locus of Control
X_3 Alienation
X_4 Social Responsibility
X_5 Machiavellianism
X_6 Political Opinion
X_7 Knowledge

To demonstrate the use of multiple regression, we will show the procedures used by HATCO to attempt to predict level of knowledge achieved by sales personnel using these six social-psychological attitude measures obtained from each employee.

The simplified procedure we will follow to demonstrate the use of multiple regression is displayed in Figure 2.5. It represents a step-by-step procedure to be followed in the application and interpretation of stepwise regression analysis. Most canned computer programs for stepwise multiple regression will automatically go through this sequence of steps.

Step 1

Table 2.6 displays all of the correlations among the six independent variables and these correlations with the dependent variable (y). Examination of the correlation matrix indicates that predictor 5 is most correlated with the dependent variable. Our first step is to build the regression equation using this 'best' predictor. Note the correlation of predictor 1 with the dependent variable is .67. However predictor 1 is correlated (.61) with predictor 5. This is your first clue that the use of both predictor variables 5 and 1 might not be appropriate, as they are

STEP NUMBER

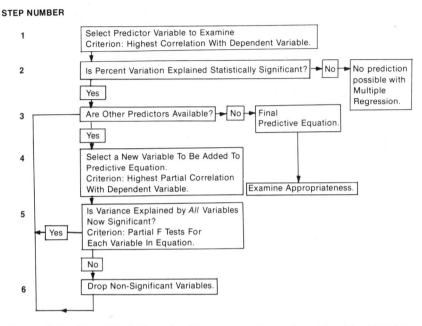

Figure 2.5—Simplified Stepwise Regression Procedure Used *by HATCO*

almost as highly correlated with each other as they are with the dependent variable. The results of this first step typically appear as shown in Table 2.7.

The concepts from Table 2.7 you should be familiar with are:

Multiple R: This value is the correlation coefficient (at this step) for the simple regression of predictor 5 and the dependent variable. It has no + or − sign because in multiple regression the signs of the

TABLE 2.6
CORRELATION MATRIX: HATCO DATA

Variables		Predictors						Dependent
		X(1)	X(2)	X(3)	X(4)	X(5)	X(6)	y
	X(1)	1.00						
Predic-	X(2)	−0.38	1.00					
tors	X(3)	0.50	−0.50	1.00				
	X(4)	0.02	0.23	−0.12	1.00			
	X(5)	0.61	0.48	0.05	0.23	1.00		
	X(6)	0.04	0.13	−0.04	0.77	0.16	1.00	
	y	0.67	0.04	0.57	0.19	0.68	0.21	1.00

TABLE 2.7
EXAMPLE OUTPUT: STEP ONE OF HATCO MULTIPLE REGRESSION EXAMPLE

STEP NO. ONE:
   VARIABLE
   ENTERED.   5 X(5)

MULTIPLE R	0.682
MULTIPLE R-SQUARE	0.466
STD. ERROR OF EST.	1.320

ANALYSIS OF VARIANCE

	SUM OF SQUARES	DF	MEAN SQUARE	F RATIO
REGRESSION	73.1	1	73.1	41.9
RESIDUAL	83.7	48	1.74	

VARIABLES IN EQUATION

VARIABLE	COEFFI-CIENT	STD. ERROR OF COEFF.	STD. REG. COEFF.	F TO RE-MOVE	VARIABLES NOT IN EQUATION	
(Y-INTERCEPT 4.19)						
X(5)	5 0.85	0.13	0.68	41.9	PAR-TIAL CORR.	F
					X(1) 0.438	11.176
					X(2) −0.445	11.635
					X(3) 0.728	53.211
					X(4) 0.039	0.072
					X(6) 0.150	1.096

individual variables may differ, so this coefficient reflects only the degree of association.

R Square: This is the correlation coefficient squared; also referred to as the coefficient of determination. This value indicates the percent of total variation in Y explained by X_5. The total sum of squares (73.1 + 83.7 = 156.8) is the squared error that would occur if we used only the mean of the dependent variable as our predictor. Using predictor X(5) reduces this error by 46.6 percent.

Standard Error Of The Estimate: This is another measure of the accuracy of our predictions. It is the square root of the sum of the squared errors divided by the degrees of freedom. Standard error =

$\sqrt{\quad}$ Sum of Squared Errors ÷ Degrees of Freedom. The standard error of the predictions is used in estimating the size of the confidence interval for the predictions. See Neter and Wasserman [4] for details regarding this procedure.

Variables in the Equation (Step 1):   The value .85 is the b_1 regression coefficient calculated from the new data. The standardized regression coefficient or beta value of .68 is the b_1 value calculated from standardized data. Note that with only one independent variable the beta coefficient equals the correlation coefficient. The beta value allows you to compare the effect of X_5 on Y to the effect on Y of other variables at each stage, since this reduces b_1 to a comparable unit, i.e., number of standard deviations. (At this step we have no other variable available for comparison).

Standard Error of Coefficient: This is the standard error of the estimate of b_1. The value of b_1 divided by the standard error (.85/.13 = 6.5) is the calculated t value for a t test of the hypothesis B = 0.

F To Remove: This is the partial F value (and in this case is identical to the overall F value). This F statistic exceeds the tabled F value for one and 48 degrees of freedom at the 99 percent confidence level. This is acceptable to HATCO's management as statistically significant.

For the section of the table headed "Variables not in Equation" an illustration will show the meaning. For the illustration, we will use variable X_3 which is not in the regression model.

Partial Correlation: This is the partial correlation of X_3 with Y given that X_5 is already in the regression model. It is an indication of the variation in Y not accounted for by X_5 that can be accounted for by X_3. Remember, the partial correlation coefficient can be misinterpreted. This does not mean that we explain 72.8 percent of the previously unexplained variance. It means that 53 percent ($72.8^2 = 53$; the partial coefficient of determination) of the *unexplained* (not the total) variance can now be accounted for. Since 47 percent was explained by X_5, $(1 - .47) \times .53 = 28.3$ percent of the total variance could be explained by adding variable X_3.

F: This column of numbers represents the partial F's for all the variables not yet in the equation. These partial F's are calculated as a ratio of the additional sum of squared errors explained by including a particular variable and the sum of squared error left after adding that same variable. If this F level is not significant at the appropriate $(1 - \alpha)$ level a variable usually would not be allowed to enter the regression model. The tabled F value at $\alpha = .01$ with 2 and 47 degrees of freedom is F = 2.09. Looking at the column of partial F's note that the F values for variables X_1, X_2, and X_3 are all larger than the tabled value. Therefore, all three variables could be considered for inclusion in the model.

Recall that the simple correlation of variable X_1 with the dependent variable (y) was 0.67 and that for variable X_3 was 0.57. Therefore, you may have thought variable X_1 would be included in the model next. But, in deciding which additional variables to include in the equation we would select first that independent variable which exhibits the highest partial correlation with the dependent variable (not the highest simple correlation). The partial correlation coefficient for X_3 is the largest (.728) and therefore X_3 will be considered for addition to the model before X_1.

We now know that a significant portion of the variance in the dependent variable is explained by predictor variable 5 (Step 2). We can also see predictor variable three has the highest partial correlation with the dependent variable, and the F ratio for this partial correlation is significant at the .01 level (Step 3). We now move to step 4 to attempt to predict using both variables 5 and 3. The output from this prediction is shown in Table 2.8.

Step 5—Variable X_3 Added

The multiple R value and R values have both increased in value. The R square has increased by the 28 percent we predicted when we examined the partial correlation coefficient for X_3 of .728. The increase in R square of 28 percent is derived by multiplying the 53.4 percent of the variation that was not explained after Step 1 by the partial F squared: $53.4 \times (.728)^2 = .28$. That is, of the 53.4 percent unexplained after Step 1, $(.728)^2$ of this variance was explained by adding variable X_3 yielding a total variance explained of $.47 + (.534 \times (.728)^2) = .749$.

Variables in the Equation

The value of b_1 has changed very little. This is a further clue that variables X_5 and X_3 are relatively independent. If the effect on y of X_3 was independent of the effect of X_5, the b_1 coefficient would not change at all.

F To Remove: The two values of F presented here are again partial F values. The F value for X_5 is now 79.6, where it was 41.9 in Step One. The F value for X_5 at this step indicates the ratio of the sum of squares due to regression added by including X_5 in the regression model as if X_3 had been in the equation first. We can therefore examine the contribution of all variables entered at earlier steps as if they were to be entered after the variable entered at this step. Note that the F value for X_3 (53.2) is the same value shown for X_3 in Step 1 under the heading "Variables Not in the Equation" (See Table 2.7).

TABLE 2.8
EXAMPLE OUTPUT: STEP FOUR (VARIABLE 3) OF HATCO
MULTIPLE REGRESSION EXAMPLE

VARIABLE ENTERED   3 X(3)

MULTIPLE R	0.865
MULTIPLE R-SQUARE	0.749
STD. ERROR OF EST.	0.914

ANALYSIS OF VARIANCE

	SUM OF SQUARES	DF	MEAN SQUARE	F RATIO
REGRESSION	117.5	2	58.78	70.35
RESIDUAL	39.2	47	0.83	

VARIABLES IN EQUATION

VARIABLE		COEFFI-CIENT	STD. ERROR OF COEFF.	STD. REG. COEFF.	F TO RE-MOVE	VARIABLES NOT IN EQUATION	
(Y-INTERCEPT		−0.855)					
X(3)	3	0.338	0.046	0.533	53.2	PAR-	
X(5)	5	0.819	0.092	0.652	79.6	TIAL	
						CORR.	F
						X(1) 0.002	0.00
						X(2) −0.006	0.00
						X(4) 0.211	2.16
						X(6) 0.279	3.89

Since both predictor 3 and 5 make significant contributions to the explanation of the variation in the dependent variable, we can go to step 3 and ask "are other predictors available?" Looking at the partial correlations for the variables not in the equation in Table 2.8 we see that predictor 6 has the highest partial correlation (.279). This variable would explain 7.7 percent of the heretofore unexplained variance $(.279)^2 = .077)$ or 2 percent of the total variance $(1 - .749) \times .077 = .02)$. This is a very modest contribution to the explanatory power of our predictive equation in spite of the significance of the F value for predictor 6 at the .05 alpha level. (Note: the tabled F value for 3 and 46 degrees of freedom at $p = .05$ is 2.81, while the F value of predictor 6 is 3.89).

TABLE 2.9
EXAMPLE OUTPUT: STEP FOUR (VARIABLE 6) HATCO
MULTIPLE REGRESSION EXAMPLE

VARIABLE ENTERED   6 X(6)

MULTIPLE R	0.877
MULTIPLE R-SQUARE	0.769
STD. ERROR OF EST.	0.887

ANALYSIS OF VARIANCE

	SUM OF SQUARES	DF	MEAN SQUARE	F RATIO
REGRESSION	120.6	3	40.2	51.08
RESIDUAL	36.2	46	0.78	

VARIABLES IN EQUATION

VARIABLE		COEFFI-CIENT	STD. ERROR OF COEFF.	STD. REG. COEFF.	F TO RE-MOVE	VARIABLES NOT IN EQUATION	
(Y-INTERCEPT	−1.684)						
X(3)	3	0.343	0.045	0.541	57.9	PAR-	
X(5)	5	0.790	0.090	0.629	76.4	TIAL	
X(6)	6	0.171	0.086	0.142	3.8	CORR.	F
						X(1) 0.017	0.013
						X(2) 0.019	0.017
						X(4) −0.004	0.001

We decide to enter predictor 6 into the regression equation and the results are shown in Table 2.9. As we predicted the value of R-square increases by two percent. In addition, examination of the partial correlations for variables 1, 2 and 4 indicates that no additional value will be gained by adding them to the predictive equation. These partial correlations are all very small and have partial F values associated with them that would not be statistically significant. We can now examine our predictive equation that includes variables 3, 5 and 6.

The section of Table 2.9 headed 'Variables in Equation' yields the prediction equation in the column labeled 'coefficient'. From this column we read the constant term in the equation or y-intercept of −1.684 and the coefficients of variables 3, 5 and 6 respectively to be .343, .790, and .171. This equation would be written:

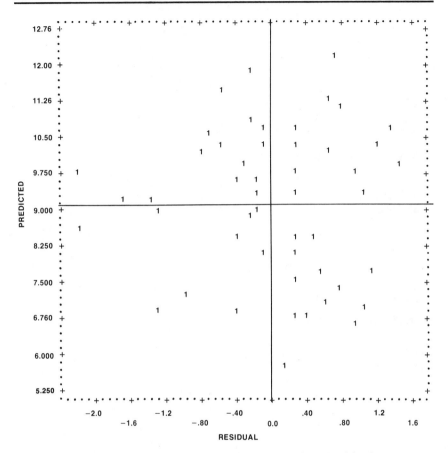

Figure 2.6—Plot Of Predicted Value Against Residuals

$$y = -1.684 + .343 X_3 + .790 X_5 + .171 X_6.$$

To examine the appropriateness of this equation we have already considered statistical significance. A look at the residual plots reinforces our judgment on the appropriateness of this predictive equation. Figure 2.6 and Figure 2.7 show a plot of the residuals against the predicted values and a plot of the normalized residuals against an expected normal distribution. The errors appear to be "reasonably" normally distributed, which gives us reassurance that the previously presented equation is appropriate.

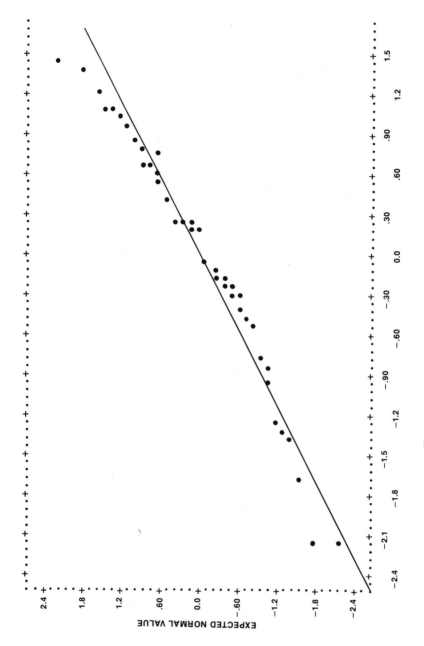

Figure 2.7—Normal Probability Plot of Residuals

A FINAL DIGRESSION: DUMMY VARIABLES

To this point all of our illustrations have implicitly assumed metric measurement for both predictor and criterion variables. The criterion variable may legitimately be measured as a dichotomous (0,1) variable. In such cases regression analysis is comparable to discriminant analysis. Using dichotomous measures as predictor variables however results in interpretational differences. The following example will help clarify this.

Assume we want to predict the number of credit cards held by families and our measurement consists of:

Age of Male head of household:
   X_1 = 1 if age under 40, 0 if 40 or older.
   X_2 = 0 if age under 40, 1 if 40 or older.

Income of household:
   X_3 = 1 if under $10,000, 0 if over.
   X_4 = 0 if under $10,000, 1 if $10–19,000, 0 if $20,000 +.
   X_5 = 0 if under $19,999, 1 if $20,000 or over.

We cannot use the least squares procedures previously discussed because we have redundant data and must "drop" some of the variables. We know that if $X_1 = 1$, then X_2 has to equal zero; so we need either X_1 or X_2, but not both. We know that if X_3 and $X_4 = 0$, then X_5 must equal 1 so we only need two of those three variables. Assume we drop X_2 and X_5. We can now use the previously discussed least squares procedure to estimate the predictive equation. The interpretation of the regression coefficients has changed, however.

To discuss the interpretation of the regression coefficient we further assume we have the following predictive equation:

$$Y = b_0 + b_1X_1 + b_3X_3 + b_4X_4.$$

If a respondent is over forty years of age and has a family income of $20,000 or over, $X_1 = 0$, $X_3 = 0$, and $X_4 = 0$; and the prediction would be $\hat{Y} = b_0$. If the respondent is under forty and has an income under $10,000, $X_1 = 1$, $X_3 = 1$, and $X_4 = 0$; and the prediction would be $Y = b_0 + b_1 + b_3$. In other words, the constant (intercept) coefficient estimates the average effect of the omitted variables, and the other coefficients represent the average differences between the omitted variables and the included variables.

SUMMARY

This chapter presents a simplified introduction to the rationale and fundamental concepts underlying multiple regression analysis. The chapter emphasizes that multiple regression analysis can describe and predict the relationship between two or more intervally scaled variables. Also, multiple regression analysis, which can be used to examine the incremental and/or total explanatory power of many variables, is a great improvement over the sequential analysis approach necessary with univariate techniques. Both stepwise and simultaneous approaches can be used to solve a regression equation, and under certain circumstances nonmetric dummy coded variables can be included in a regression equation. This chapter will give you a fundamental understanding of "how regression works" and what can be achieved through its use. Also, familiarity with the concepts presented in this chapter will help you better understand the more complex and detailed technical presentations in other texts.

END OF CHAPTER QUESTIONS

1. How would you explain the "relative importance" of the predictor variables used in a regression equation?
2. Why is it important to examine the assumption of linearity when using regression?
3. Do you think you could find a regression equation that would be acceptable as "statistically significant" and yet offer no acceptable interpretational value to management? How could this happen?
4. What is the difference in interpretation between the regression coefficients associated with interval scaled predictor variables as opposed to dummy (0,1) predictor variables?
5. What is the difference between interactive predictor variables and correlated predictor variables? Do any of these differences affect your interpretation of the regression equation?

REFERENCES

1. Draper, Norman and Harry Smith, *Applied Regression Analysis*, John Wiley & Sons, 1966.
2. Huang, David S., *Regression and Econometric Methods*, John Wiley & Sons, Inc., 1970.
3. Johnston, J., *Econometric Methods*, (2nd ed.) McGraw-Hill Book Company (1972).

4. Neter, John and William Wasserman, *Applied Linear Statistical Models,* Richard D. Irwin, Inc., 1974.
5. Nie, Norman, Dale Bent, and Hadlai Hull, *Statistical Package for the Social Sciences,* McGraw-Hill Book Company (1970).

SELECTED READINGS

MEASURING SALES EFFECTS OF SOME MARKETING MIX VARIABLES AND THEIR INTERACTIONS

V. KANTI PRASAD and L. WINSTON RING*

Introduction

Effective marketing management requires information about the relative sales effectiveness of different marketing variables and also their interactive effects. Yet, too often, only time-aggregated data on single variables are collected in simple research designs such as before and after an advertising campaign. It is then uncertain which of the marketing mix variables are more important and how these variables interact in affecting sales. More elaborate controlled field experiments generally are regarded as analytically very desirable, but too expensive or too unwieldy, and hence such experiments are rarely used.

The Milwaukee Advertising Laboratory and an anonymous manufacturer provided longitudinal data from a unique, well planned and controlled field experiment involving two aggregate levels of TV advertising. The analysis herein focuses upon how price and several forms of advertising interact in affecting market shares, and whether these effects are modulated by the aggregate level of TV advertising.

Design

The primary data were obtained from the consumer panels of the Milwaukee Advertising Laboratory, a now discontinued subsidiary of The Journal Company of Milwaukee. Three features of their research

* V. Kanti Prasad and L. Winston Ring are Associate Professors at the School of Business Administration of the University of Wisconsin-Milwaukee.

The writers are grateful to The Journal Company, the Milwaukee Advertising Laboratory, and an anonymous manufacturer for providing the data for this research. A research assistant, Renald Paberz, extracted and organized the data from a massive nonstandard data bank. Another research assistant, Imran Currim, performed the many computer calculations and provided valuable comments. The research was supported by the Graduate School of the University of Wisconsin-Milwaukee.

Reprinted by permission of the authors and publisher from the Journal of Marketing Research, *published by the American Marketing Association Vol. 13 (November 1976) pp. 391–396.*

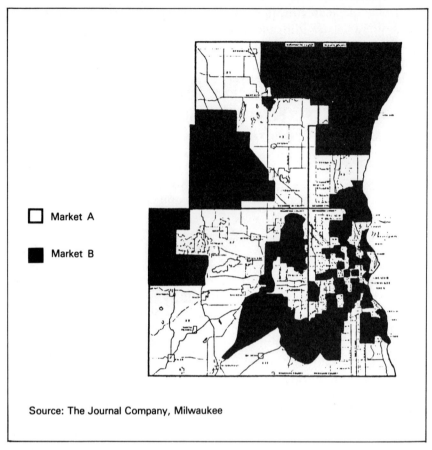

Source: The Journal Company, Milwaukee

Figure 1—*Matched Markets of the Milwaukee Advertising Laboratory*

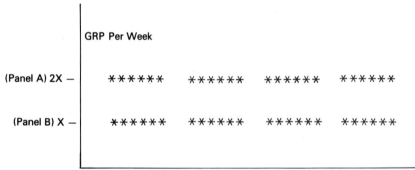

Figure 2—*Blocking Chart of Television Inputs for the Two-Level Advertising Test in The Milwaukee Advertising Laboratory*

facilities used for this project were (1) matched markets, (2) separate consumer panels in the matched markets, and (3) divided runs on television commercials.

The geographic division of the Milwaukee area into markets A and B is shown in Figure 1. The areas were designed so that the respective total populations were matched on major demographic variables such as income, home ownership, family size, race, and religion.

The consumer panels of approximately 750 families each in the A and B markets also matched on these demographic variables. These families reported weekly the purchase details of specified product classes.

Differentiated TV advertising to the A and B panels was made possible by a special muting device attached (with permission) to the TV sets of panel B members. When triggered by a certain high frequency signal from any of the three major TV stations (ABC, CBS, NBC), the muting device supressed the audio and visual output of the TV set. A second signal restored the sound and the picture. This method was used to prevent panel B members from viewing some selected TV ads, while panel A members had full opportunity to be exposed to the advertisements. The data from the two sets of purchase diaries thus provide a direct means of determining the effects on purchasing of the two controlled levels of TV advertising exposure opportunity in panels A and B.

A nationally known manufacturer of consumer goods used the panels for a 64-week test on TV advertising for one of its brands. The product was a frequently purchased canned food item priced at about 60¢ for the most commonly purchased size.

The design for delivery of TV advertising to both panels consisted of four flights of about six weeks each with intervals from four to 13 weeks between. As shown in Figure 2, the television advertising inputs to panel A were almost exactly twice those to panel B.

The difference in Gross Rating Points between the two panels was accomplished by use of spot TV. In each of the four flights, network advertising was used to the extent of X units. The panels had equal opportunity for exposure to these ads. Spot TV of X units per week was used in a similar time schedule, but the muting devices were used to prevent panel B families from seeing any of these ads. The spot and network ads were produced by the same New York agency, and both consisted of a mix of two versions on the same theme.

The model

A linear model was hypothesized for each experimental panel with the experimental brand's market share as the dependent variable.

A separate set of coefficients was permitted for each panel because of the possibility that the higher aggregate level of TV advertising input to the one panel may have altered the pattern of response to the marketing mix variables. Market shares were measured over 64 successive weekly time periods. As shown by Sexton [9], aggregation over longer time spans can obscure meaningful relationships. The use of a linear model is consistent with the work of other researchers, and it was determined by Sexton [10] that his theoretically derived nonlinear formulations did not perform better empirically.

The set of independent variables included lagged market share, and concurrent and lagged measures of relative price, TV advertising, retailer advertising in newspapers, and magazine advertising. Also, as suggested by Kotler [6, p. 72–9], interactions were included as pairwise multiplicative variables. Competitors' variables included were magazine and newspaper advertising. The full model is:

$$
\begin{aligned}
MS_t = \ & \beta_1 + \beta_2 MS_{t-1} + \beta_3 P_t + \beta_4 P_{t-1} + \beta_5 TV_t \\
& + \beta_6 TV_{t-1} + \beta_7 NEW_t + \beta_8 NEW_{t-1} \\
& + \beta_9 MAG_t + \beta_{10} MAG_{t-1} + \beta_{11} P_t{}^* TV_t \\
& + \beta_{12} P_t{}^* NEW_t + \beta_{13} TV_t{}^* NEW_t \\
& + \beta_{14} P_t{}^* MAG_t \\
& + \beta_{15} TV_t{}^* MAG_t + \beta_{16} NEW_t{}^* MAG_t \\
& + \beta_{17} MAG_t^o + \beta_{18} NEW_t^o + e_t,
\end{aligned}
$$

where the superscript o refers to all other competitors.

Complete definitions of the operational measures are given in Appendix A. Note that P_t was measured as the price of the experimental brand in relation to the weighted prices of the other brands in period t. Weiss [11] found empirically that this definition performs much better than raw price. Limited availability of data prevented similar definitions of some advertising variables; however, Frank and Massey [4] found that characterization of retail advertising as an index adds little toward predictive efficacy.[1]

---

[1] Limited dependent variables such as market share pose some theoretical difficulties in model formulation and in prediction [8]. Limited rather than unrestricted ranges for the explanatory variables are theoretically more desirable in such situations. However, in the present research, all predicted market shares were in the proper 0–1 range, and, in fact, were more than four standard deviations from either boundary. For this reason, it is believed that the problem of boundedness of the dependent variable does not have any practical consequences in interpretation of the results, although extrapolation much beyond the ranges of observed values is hazardous.

Analysis

Two approaches were used in analyzing the data in each panel. Basic to both approaches is the assumption that the two panels are comparable except for the effects of the experimental variable of TV advertising. This comparability seems assured by the careful demographic matching of the two panels; moreover, an analysis by an independent researcher for the Milwaukee Advertising Laboratory showed the market shares of the experimental brand to be not significantly different in the two panels immediately before the experiment. Both approaches fitted the model to the data by the method of least squares. The writers recognize that because the dependent variable is a proportion, the assumption of homoscedasticity may not be tenable. However, when generalized least squares estimators were computed under the assumption that the error variances are proportional to $p(1 - p) / n$ in a manner suggested by Goldberger [5, p. 235], it became obvious that these weights dominated the explanatory variables in the analysis and introduced high spurious correlations. Hence this approach was rejected. The arcsin transformation [1] also was used as a method of stabilizing the error variance. The subsequent regression results were virtually the same as those obtained with untransformed current and lagged market shares. For reasons of simplicity and ease of interpretation, only the results for untransformed variables are presented.

Stepup Approach.   In this first approach, the coefficients were estimated for each panel by a stepup regression method. Successive individual explanatory variables were included in the model by use of a sequential selection criterion of maximum partial correlation with the dependent variable. This is one heuristic method of determining an order of dominance of individual effects. Table 1 presents the final coefficients in the two panels significant at a probability level of .15 or less.

Table 1 shows that in panel A with the high television input, relative price is the single most useful variable in predicting market share. Price is also useful for prediction of market share in the low TV panel, but it is lower in the hierarchy of inclusion and the coefficient is much smaller. Thus, a frequent expectation that higher advertising levels will support higher prices [7] is not supported for this brand.

In panel B with the low TV advertising input, a noticeable finding is that the dominant single determinant of current market share is the previous period's market share. This suggestion of the presence of carryover effects in the low TV panel is explored further in the following section.

TABLE 1
STEPWISE ESTIMATION BY INCLUSION

	Order	Variable	Final regression coefficients	Level of significance	R^2	Adj. R^2
Panel A		Constant	1.31	.000		
(high TV	1st	P_t	−1.000	.000	.34	.32
advertising	2nd	MAG^0_{t-1}	−.0000019	.036	.37	.35
input)	3rd	TV_{t-1}	.000151	.086	.40	.37
		Constant	.132	.000		
Panel B	1st	MS_{t-1}	.325	.003	.21	.20
(low TV	2nd	NEW_t	.00125	.006	.34	.31
advertising	3rd	P_t	−.378	.016	.39	.36
input)	4th	NEW^0_t	−.00123	.036	.42	.38
	5th	$TV_t * MAG_t$	.000060	.125	.44	.39

From both panels there is evidence that the direct effects of TV advertising may generally be difficult to isolate. In neither panel does TV_t itself appear as a significant variable. Also, the sets of variables most useful in predicting market share are different for the two panels, and this finding is hypothesized to be due to the one known difference between the two panels—different levels of TV advertising. Thus, aggregate TV advertising appears to be operating as a modulating variable on the order of dominance of the other individual marketing mix variables.

Stepdown Approach. This method of analysis also used stepwise regression, but in a stepdown or exclusion version. The method first incorporated the full set of explanatory variables. Then, successive variables were deleted whose partial correlations with MS_t were smallest if the corresponding coefficient was not significant for $\alpha = .15$. At each stage, the set of previously deleted variables was re-evaluated to determine whether individual variables should be reincorporated in the model. This analysis was performed separately for each panel. For direct comparison purposes, the model then was refitted to the data for each panel by use of the set of variables whose coefficients were significant in either panel. The results are shown in Table 2. Though this stepwise procedure does not show the relative dominance of individual effects as does the first procedure, it minimizes the risk of omitting relevant variables and thus avoids producing biased estimates [3, p. 389].

TABLE 2
STEPWISE ESTIMATION BY EXCLUSION

Variable	Final regression coefficients		Level of significance	
	Panel A (high)	Panel B (low)	Panel A (high)	Panel B (low)
Constant	.239	.156	.000	.000
MS_{t-1}	−.0250	.310	.800	.008
TV_t	.00686	.00819	.000	.028
NEW_t	.0422	.0172	.000	.185
$P_t{}^*TV_t$	−.00632	−.00722	.000	.033
$P_t{}^*NEW_t$	−.0387	−.0144	.000	.247
$P_t{}^*MAG_t$	.00354	.00162	.031	.370
$NEW_t{}^*TV_t$	−.00000993	−.0000233	.077	.068
$NEW_t{}^*MAG_t$	−.000201	−.000121	.126	.432
NEW_t^0	.0000749	−.00114	.881	.045
MAG_{t-1}^0	−.00000138	−.00000075	.081	.403
R^2	.60	.48	.000	.000
Durbin-Watson statistic	1.86	1.75		

Table 2 also presents the informational content of the full set of variables in a very parsimonious manner. More than half of the variables have been deleted, but the respective R^2's only decrease from .61 to .60 and from .51 to .48. By use of an F-test [5, p. 175], the coefficients of the deleted variables can be assumed to be zero at an α of .01.

Pairwise comparisons of the two sets of regression coefficients suggest differences between panels in the carryover effects of the marketing mix variables. In the high advertising input panel, the immediate effects of marketing variables appear to be dominant, whereas in the lower advertising input panel, the presence of significant delayed or carryover effects of prior marketing variables is indicated by the large positive coefficient for the lagged market share variable. However, Chow's [2] test of the hypothesis that the full set of coefficients is the same for both panels cannot be rejected at a reasonable level of significance.

The writers tested a more focused hypothesis that for sufficiently high levels of TV advertising, the influence of TV advertising on market share would decrease and the influence of the other promotional mix variables would increase.[2] Translated in terms of variables used in the

---

[2] This hypothesis was developed earlier on the basis of discussions with Joseph Chamberlin of Million Market Newspapers in New York.

final model in Table 2, this hypothesis states that in the high TV advertising panel A, the absolute values of the coefficients for TV_t and $P_t * TV_t$ should be less than in panel B, and the absolute values of coefficients for NEW_t, $P_t * NEW_t$, $P_t * MAG_t$, and $NEW_t * MAG_t$ should be greater than in the low TV panel B. The coefficient of MAG_t is not tested because it has no significant single effect on market share, and $NEW_t * TV_t$ is not included because the hypothesized effects of the two variables involved are opposing. Because the hypothesis is directional only, a sign test is used. All six pairs of coefficients are in accord with the expected directional pattern, and the hypothesis is upheld with a significance level of from $(1/2)^3 = .125$ to $(1/2)^6 = .015$, depending upon the extent of independence of the six events.

The results in Table 2 show also that interactions among the marketing mix variables are statistically significant in determining market share. In the stepup results, price was individually most useful in one panel, and somewhat less useful in the other panel; yet in both panels the individual effect of price is subsumed by interactive effects when these interaction variables are explicitly included.

The significant effect of price in the case of the brand studied appears to be in how it interacts with TV, newspaper, and magazine advertising. These interactions can determine whether a marketing mix variable has a favorable or an unfavorable influence upon a brand's market share. For example, the change of market share in response to a marginal change in TV advertising in panel A is:

$\partial(MS_t)/\partial(TV_t) = .00686 - .00632P_t - .00000993NEW_t.$

This shows that the effect on market share of increased TV advertising is modulated by the brand's relative price and its newspaper advertising. Further, it can be shown by evaluation at the average value for NEW_t that if the brand's price is only 7% higher than the competitor's average price, the effect of increased TV advertising is a decreased market share.

Although causal inference is hazardous at best, one plausible explanation for the observed pattern of interaction effects between the advertising and price variables may lie in the message content of the TV advertisements for the brand used during the analysis period. The TV advertising copy employed during this period focused on the theme of versatile use of the product class in question, but did not emphasize the "superior quality" of the brand in relation to competitive brands. At the same time, newspaper advertising by retailers focused upon the price of the brand. Without the perception of superior quality for the brand and faced with an actual high relative price for it on the supermarket shelves, consumers might have shown a tendency to choose a rival brand which was cheaper or whose advertising projected a higher quality image.

SUMMARY

Data obtained from a carefully controlled field experiment conducted by the Milwaukee Advertising Laboratory were analyzed by use of a linear model to explain variations in market share. It was found that the effects of TV, newspaper, and magazine advertising for the experimental brand were dependent upon the relative price level of the brand. Significant interaction effects between newspaper, TV, and magazine advertising also were identified. The results indicated that immediate effects of marketing variables were dominant in the market which received the higher aggregate level of TV advertising input whereas lagged or carryover effects were significant in the market which received the lower level of TV advertising. A hypothesis of decreasing returns to scale for TV advertising with a concurrent shifting of importance to the other promotional mix variables was examined and tentatively maintained.

The revealed complexity of the mechanism of response to marketing mix variables indicates the usefulness of more such field experiments in marketing research. The present study also reaffirms the need to consider the interactive effects of marketing mix variables in building realistic decision models in marketing.

APPENDIX A

DEFINITIONS OF VARIABLES

MS_t The ounces of the experimental brand reported purchased in week t in relation to the total reported purchases of the product.

P_t The reported average price per ounce of the experimental brand in relation to the reported average price of all other brands.

TV_t Network and spot television advertising for the experimental brand measured in G.R.P. as reported by the advertising agency.

NEW_t Newspaper advertising for the brand by retailers in week t. (Computed as a summation of the number of times supermarket chains featured the brand in newspaper ads weighted by the market share of the corresponding chain.)

MAG_t Magazine advertising for the brand by the manufacturer in week t. (Computed as a summation of the pages or fractions of pages of ads in 14 leading women's magazines weighted by the relative readership. The readership rates for adult women were obtained from the B.R.I. Basic Magazine Report. Monthly issues were proportioned over a forward four-week period by weights of 1, 1, 1/2, 1/2.)

REFERENCES

1. Brownlee, K. A. *Statistical Theory and Methodology in Science and Engineering.* New York: John Wiley & Sons, 1965.
2. Chow, Gregory C. "Test of Equality Between Sets of Coefficients in Two Linear Regressions," *Econometrica*, 28 (July 1960), 591–605.
3. Christ, Carl F. *Econometric Models and Methods.* New York: John Wiley & Sons, 1966.
4. Frank, Ronald E. and William F. Massy. "Effects of Short-Term Promotional Strategy in Selected Market Segments," in Patrick J. Robinson, et al., *Promotional Decisions Using Mathematical Models.* Boston: Allyn and Bacon, 1967, 149–225.
5. Goldberger, Arthur S. *Econometric Theory.* New York: John Wiley & Sons, 1964.
6. Kotler, Philip. *Marketing Decision Making: A Model Building Approach.* New York: Holt, Rinehart, and Winston, 1971.
7. Kuehn, Alfred A. "How Advertising Performance Depends on Other Marketing Factors," *Journal of Advertising Research*, 2 (March 1962), 2–10.
8. McGuire, T. W., J. U. Farley, R. E. Lucas, and L. W. Ring. "Estimation and Inference for Linear Models in Which Subsets of the Dependent Variable are Constrained," *Journal of the American Statistical Association*, 63 (December 1968), 1201–13.
9. Sexton, Donald. "Estimating Marketing Policy Effects on Sales of a Frequently Purchased Product," *Journal of Marketing Research*, 7 (August 1970), 338–47.
10. _____. "A Microeconomic Model of the Effects of Advertising," *Journal of Business*, 45 (January 1972), 29–41.
11. Weiss, Doyle. "The Determinants of Market Share," *Journal of Marketing Research*, 5 (August 1968), 290–5.

3

Multiple Discriminant Analysis

CHAPTER REVIEW

Much has been written on the multivariate statistical technique of multiple discriminant analysis. This chapter understandably discusses an otherwise complex and sophisticated technique—multiple discriminant analysis, without resorting to statistical "jargon" and mathematical formulas or limited explanation of important concepts. The chapter has two objectives: (1) to introduce the underlying nature, philosophy, and conditions of discriminant analysis; and (2) demonstrate its application and interpretation with an illustrative example. Before proceeding further, it will be helpful to define the Key Terms.

Students who understand the most important concepts in the area of discriminant analysis should be able to:

☐ State the circumstances under which a linear discriminant function rather than multiple regression should be used.

☐ Identify and compare the three major stages in the application of discriminant analysis.

☐ Identify the two computational approaches for discriminant analysis and state when each should be used.

☐ Tell how to interpret the nature of a linear discriminant function, i.e. identify independent variables with significant discriminatory power.

☐ Explain the usefulness of the classification matrix methodology.

☐ Describe how to develop a classification matrix.

☐ Explain a chance model.

☐ Differentiate between the hit-ratio and multiple regression's R^2.

☐ Justify the use of a split sample approach to validating the discriminant function.

DEFINITIONS OF KEY TERMS

CRITERION VARIABLE.   Dependent variable.

PREDICTOR VARIABLE.   Independent variable.

CATEGORICAL VARIABLE.   Referred to by some as a nonmetric, nominal, binary, qualitative, or taxonomic variable. When a number or value is assigned to a categorical variable it serves merely as a label or means of identification. The number on a football jersey is an example.

METRIC VARIABLE.   A variable with a constant unit of measurement. If a variable is scaled from 1 to 9, the difference between 1 and 2 is the same as that between 8 and 9.

LINEAR COMBINATION.   Also referred to as linear composites, linear compounds, and discriminant variates, they represent the weighted sum of two or more variables.

DISCRIMINANT FUNCTION.   A linear equation in the following form:

$$Z = W_1X_1 + W_2X_2 + \ldots + W_nX_n$$

where

$$Z = \text{the discriminant score}$$
$$W = \text{discriminant weight}$$
$$X = \text{independent variable}$$

DISCRIMINANT SCORE.   Referred to as a Z-score; defined by the previous equation.

DISCRIMINANT WEIGHT.   Referred to by some as a discriminant co-efficient, its size is determined by the variance structure of the original variables. Independent variables with large discriminatory power usually have large weights and those with little discrim-inatory power usually have small weights; collinearity among the independent variables will cause exception to this rule.

CENTROID.   The mean value for the discriminant Z-scores for a particular category or group. A two-group discriminant analysis has two centroids, one for each of the groups.

DISCRIMINANT LOADINGS.   Referred to by some as structure cor-relations, they measure the simple linear correlation between the independent variables and the discriminant function.

ANALYSIS SAMPLE.   When constructing classification matrices, the original sample should be divided randomly into two groups. One for developing the discriminant function and the second for validating it. The group used to compute the discriminant function is referred to as the analysis sample.

HOLD-OUT SAMPLE.   Also referred to as the validation sample, it is the group of subjects "held-out" of the total sample when the function is computed.

CLASSIFICATION MATRIX.   Also referred to as a confusion, assign-ment, or prediction matrix. It is a matrix containing numbers which reveal the predictive ability of the discriminant function. The numbers on the diagonal of the matrix represent correct classifications and the off diagonal numbers are incorrect classi-fications.

CUTTING SCORE.   The criterion (score) against which each indi-vidual's discriminant score is judged to determine into which group the individual should be classified. When the analysis involves two groups, the hit ratio is determined by computing a single "cutting" score. Those entities whose Z-scores are below this score are assigned to one group while those above are classified in the other group.

HIT RATIO. The percentage of statistical units (individuals, respondents, objects, etc.) correctly classified by the discriminant function.

3

WHAT IS DISCRIMINANT ANALYSIS?

In attempting to select an appropriate analytical technique, we sometimes encounter a problem which involves a categorical dependent variable and several metric independent variables. For example, we may wish to identify individuals who are "good" credit risks from those who are "bad" risks, based on family income and size.

Discriminant analysis is the appropriate statistical technique when the dependent variable is categorical (nominal or nonmetric) and the independent variables are metric. In many cases the dependent variable will consist of two groups or classifications. For example, male versus female or high versus low. In other instances more than two groups are involved, such as a three group classification involving low, medium, and high classifications. Discriminant analysis is capable of handling either two groups or multiple groups (three or more). When two classifications are involved the technique is referred to as two-group discriminant analysis. When three or more classifications are identified the technique is referred to as multiple discriminant analysis (MDA).

Discriminant analysis involves deriving the linear combination of the two (or more) independent variables that will discriminate best between the a priori defined groups. This is achieved by the statistical decision rule of maximizing the between-group variance relative to the within-group variance—this relationship is expressed as the ratio of the between-group to within-group variance. The linear combinations for a discriminant analysis are derived from an equation which takes the following form:

$$Z = W_1X_1 + W_2X_2 + W_3X_3 + \ldots + W_nX_n$$
where
Z = the discriminant score
W = the discriminant weights
X = the independent variables

Discriminant analysis is the appropriate statistical technique for testing the hypothesis that the group means of the two or more groups are equal. To do so, discriminant analysis multiplies each independent variable by its corresponding weight and adds these products together (see above equation). The result is a single composite discriminant score for each individual in the analysis. By averaging the discriminant scores for all the individuals within a particular group, we arrive at the group mean. This group mean is referred to as a *centroid*. When the analysis involves two groups there are two centroids; with three groups there are three centroids, and so forth. The centroids indicate the most typical location of an individual from a particular group, and a comparison of the group centroids tells how far apart the groups are along the dimension being tested.

The test for the statistical significance of the discriminant function is a generalized measure of the distance between the group centroids. It is computed by comparing the distribution of the discriminant scores for the two or more groups. If the overlap in the distributions is small, the discriminant function separates the groups well. If the overlap is large the function is a poor discriminator between the groups. The distributions of discriminant scores shown in Figure 3.1 further illustrate this concept. For example, the top diagram represents a function that separates the groups well, whereas the lower diagram shows a function that is a relatively poor discriminator between groups A and B (note that the shaded areas represent probabilities of misclassifying statistical units from A and B).

Analogy with Regression and ANOVA

The application and interpretation of discriminant analysis is much the same as in regression analysis. That is, a linear combination of metric measurements for two or more independent variables is used to describe or predict the behavior of a single dependent variable. The key difference is that discriminant analysis is appropriate for research problems in which the dependent variable is categorical (nominal or nonmetric), whereas in regression the dependent variable is metric.

Discriminant analysis is also comparable to analysis of variance (ANOVA). In discriminant analysis the single dependent variable is categorical and the independent variables are metric. The opposite is true of ANOVA. ANOVA involves metric dependent variables and a single categorical independent variable.

ASSUMPTIONS OF DISCRIMINANT ANALYSIS

It is desirable that certain conditions be met for proper application of discriminant analysis. The assumptions for deriving the discriminant function are multivariate normality of the distributions and unknown (but equal) dispersion and covariance structures for the groups. When classification accuracies are determined we also must

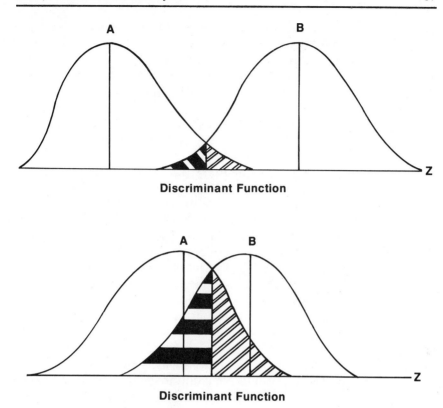

Discriminant Function

Discriminant Function

Figure 3.1—Univariate Representation Of Discriminant Z-Scores

assume equal costs of misclassification, equal a priori group probabilities, and known dispersion and covariation structures. There is evidence, however, that discriminant analysis is not very sensitive to violations of these assumptions unless the violations are extreme [7]. This is particularly true with large sample sizes.

HYPOTHETICAL EXAMPLE OF DISCRIMINANT ANALYSIS

Let's refer to a HATCO research problem to demonstrate the underlying logic of MDA. Suppose HATCO wants to find out whether one of its new products—a new and improved food mixer—will be commercially acceptable. In carrying out the investigation, HATCO is primarily interested in identifying (if possible) those consumers who "would purchase" the new product and those who "would not purchase" the new product. In statistical terminology, then, HATCO would like to minimize the number of errors they would make in deciding which consumers would buy the new food mixer and which would not. To assist in identifying potential purchasers, HATCO has devised three rating scales to be used by consumers to evaluate the new product. Consumers would evaluate the durability, performance, and style of the new product with the three scales. Rather than relying on each scale as a separate measure, HATCO believes a weighted combination of the three scales would better predict whether a consumer is likely to purchase the new product.

The statistical technique multiple discriminant analysis can be utilized to obtain a weighted combination of the three scales. Then, this weighted combination can be used to predict the likelihood of a consumer purchasing the product.

In addition to determining whether persons who are likely to purchase the new product can be identified from those who would not, HATCO would like to know which characteristics of its new product are useful in differentiating purchasers from non-purchasers. That is, which of the three characteristics of the new product—durability, performance, or style—best separates purchasers from non-purchasers. For example, if the response "would purchase" the new mixer is always associated with a high rating for durability and the response "would not purchase" is always associated with a low durability rating, HATCO could conclude that durability is a charac- teristic that discriminates well in separating purchasers from non- purchasers. In contrast, if HATCO found that about as many persons with a high rating on style said they would purchase as said they would not, then "style" would be a characteristic that discriminates poorly between purchasers and nonpurchasers.

In Table 3.1, we assume the ratings represent the judgements of a panel of 10 housewives who are potential purchasers of the new mixer, and that a particular price was specified for the product (in rating the product, each housewife would be implicitly comparing them with products already on the market). Also, after the product was evaluated by the housewives they were asked to complete a buying intention scale. From the results of this question five respondents were classified in the "would purchase" group and five as "would not purchase" the new product. As noted from Table 3.1 the difference between mean ratings for "would purchase" and "would not pur- chase" on the characteristic "durability" is high (8.0 − 3.2 = 4.8). Thus, durability appears to discriminate well between the "would purchase" and "would not purchase" groups, and is likely to be an

TABLE 3.1
HATCO SURVEY RESULTS FOR EVALUATION
OF NEW CONSUMER PRODUCT

| Purchase
Intention | Subject
Number | Ratings on Characteristics
(0 = Very Poor; 10 = Excellent) | | |
		X_1 Durability	X_2 Performance	X_3 Style
"Would	1	8	9	6
Purchase"	2	6	7	5
	3	10	6	3
	4	9	4	4
	5	7	8	2
Mean Rating		8.0	6.8	4.0
"Would Not	6	5	4	7
Purchase"	7	3	7	2
	8	4	5	5
	9	2	4	3
	10	2	2	2
Mean Rating		3.2	4.4	3.8
Difference Between Mean Ratings		4.8	2.4	0.2

important characteristic to potential purchasers.[1] On the other hand, the characteristic "style" has a difference between mean ratings for the "would purchase" and "would not purchase" groups of only 0.2 (4.0 − 3.8 − 0.2). Therefore, we would expect this characteristic to be less discriminating in terms of a decision to purchase or not to purchase the new food mixer.

The multiple discriminant analysis technique follows a procedure very similar to that shown in the hypothetical example. It identifies the areas (characteristics) where the greatest difference exists between the groups; derives a discriminant weighting coefficient for each variable to reflect these differences, and then assigns each individual to a

---

[1] This conclusion is based only on differences in the means and could possibly change after consideration of the standard deviations of the two sets of data. That is, with large standard deviations the difference between the means may not be statistically significant.

group using the weights and each individual's ratings on the characteristics.

OBJECTIVES OF DISCRIMINANT ANALYSIS

A review of the objectives for applying discriminant analysis should further clarify its nature. These include:
  (1) Determining if statistically significant differences exist between the average score profiles of the two (or more) a priori defined groups.
  (2) Establishing procedures for classifying statistical units (individuals or objects) into groups on the basis of their scores on several variables.
  (3) Determining which of the independent variables account most for the differences in the average score profiles of the two or more groups [6].
As can be noted from the above objectives, discriminant analysis is useful when the analyst is interested either in understanding group differences or in correctly classifying statistical units into groups or classes. Discriminant analysis, therefore, can be considered either a type of profile analysis or an analytical predictive technique. In either case, the technique is most appropriate where there is a single, categorical dependent variable and several metrically scaled independent variables.

GEOMETRIC REPRESENTATION

A graphical illustration of a two-group analysis will help to further explain the nature of discriminant analysis [6]. Figure 3.2 represents a scatter diagram and projection which shows what happens when a two-group discriminant function is computed. Let's assume we have two groups, A and B, and two measurements, X_1 and X_2 on each member of the two groups. We can plot in a scatter diagram the association of variable X_1 with variable X_2 for each member of the two groups. Group membership is identified by the use of X's and circles. In Figure 3.2 the X's represent the variable measurements for the members of group A and the circles are the variable measurements for group B. The ellipses drawn around the X's and circles would enclose some prespecified proportion of the points, usually 95 percent or more in each group. If we draw a straight line through the two points where the ellipses intersect and then project the line to a new axis Z, we can

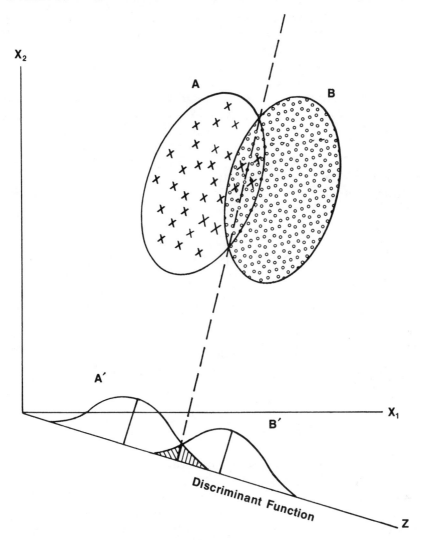

Figure 3.2—Graphical Illustration of Two-Group Discriminant Analysis

say that the overlap between the univariate distributions A′ and B′ (represented by the shaded area) is smaller than would be obtained by any other line drawn through the ellipses formed by the scatter plots [6].

The important thing to note about Figure 3.2 is that the Z axis expresses the two-variable profiles of groups A and B as single numbers (discriminant scores). By finding a linear combination of the original variables X_1 and X_2 we can project the result as a discriminant function. For example, if the crosses and circles are projected onto the new Z axis as discriminant Z-scores, the result condenses the information about group differences (shown in the X_1X_2 plot) into a set of points (Z-scores) on a single axis.

To summarize, for a given discriminant analysis problem the computer derives a linear combination of the independent variables. The result is a series of discriminant scores for each individual in each group. The discriminant scores are computed based on the statistical rule of maximizing the variance between the groups and minimizing the variance within the groups. If the variance between the groups is large relative to the variance within the groups, we say that the discriminant function separates the groups well.

APPLICATION OF DISCRIMINANT ANALYSIS

The application of discriminant analysis can be divided into three major stages: (1) derivation, (2) validation, and (3) interpretation. The *derivation stage* involves determining whether or not a statistically significant function can be derived to separate the two (or more) groups. The *validation stage* involves developing a classification matrix to further evaluate the predictive accuracy of the discriminant function. The *interpretation* stage involves determining which of the independent variables contribute the most toward discriminating between the groups. Each of these stages will be discussed in the following sections.

Stage One: Derivation

The derivation stage consists of several separate steps. The steps are: variable selection, sample division, computational method, and statistical significance.

Variable selection. To apply discriminant analysis, the analyst first must specify which variables are to be independent variables and which is to be the dependent variable. Recall that the dependent variable is categorical and the independent variables are metric.

The analyst should focus on the dependent variable first. The number of dependent variable groups (categories) can be two or more, but these groups must be mutually exclusive and exhaustive. In some cases the dependent variable may be two groups (dichotomous) such as good versus bad. In other cases the dependent variable may involve several groups (multichotomous), such as predicting occupation— i.e., physician, attorney, or professor.

The preceding examples of categorical variables were true dichotomies (or multichotomies). There are some situations, however, where discriminant analysis is appropriate even if the dependent variable is not a true categorical variable. We may have a dependent variable which is of ordinal or interval measurement which we wish to use as a categorical dependent variable. In such cases we would have to create an artificial dichotomy. For example, if we had a variable which measured the average number of colas consumed per day, and the individuals responded on a scale from zero to eight or more per day, then we could create an artificial trichotomy (three groups) by simply designating those individuals who consumed 0, 1 or 2 cokes per day as light users, those who consumed 3, 4 or 5 per day as medium users, and those who consumed 6, 7, 8 or more as heavy users of cola. Such a procedure would create a three group categorical variable in which the objective would be to discriminate between light, medium, and heavy users of colas.

Any number of artificial categorical groups can be developed. Most frequently the approach would involve creating two, three, or four categories. But a larger number of categories could be established if the need arose. When three or more categories are created, the possibility arises of examining only the extreme groups in a two group discriminant analysis. This procedure is called the polar-extremes approach.

The *polar extremes* approach involves comparing only the extreme two groups and excluding the middle group (or groups) from the discriminant analysis. The analyst could examine the light and heavy users of cola drinks and exclude the medium users. This approach can be used any time the analyst wishes to examine only the extreme groups. However, the analyst may want to try this approach when the results of a regression analysis are not as good as anticipated. Such a procedure may be helpful because it is possible that group differences may appear even though regression results were poor. That is, the polar extremes approach with discriminant analysis can reveal differences that are not as prominent in a regression analysis of the full data set [6]. Such manipulation of the data naturally would necessitate caution in interpreting one's findings.

After a decision has been made on the dependent variable, the analyst must decide which independent variables to include in the analysis. Independent variables usually are selected in two ways. The first approach involves identifying variables either from previous research or from the theoretical model which is the underlying basis of the research question. The second approach is an intuitive one. It basically involves trying to extend the researcher's knowledge and intuitively selecting variables for which no previous research or theory exists, but which logically might be related to predicting the groups for the dependent variable.

Sample Division. When applying discriminant analysis, the analyst will want to test the validity of the discriminant function which has been derived. A number of procedures have been suggested for doing this, but the most popular one involves developing the discriminant function on one group and then testing it on a second group. The usual procedure is to randomly divide the total sample of respondents into two groups. One of these groups, referred to as the *analysis sample,* is used to develop the discriminant function. The second group, referred to as the *holdout sample,* is used to test the discriminant function. This method of validating the function is referred to as the split-sample or cross-validation approach [5].

The justification for dividing the total sample into two groups is that an upward bias will occur in the prediction accuracy of the discriminant function if the individuals used in developing the classification matrix are the same as those used in computing the function. That is, the classification accuracy will be higher than is valid for the discriminant function if it were used to classify a separate sample. The implications of this upward bias are particularly important when the analyst is concerned with the external validity of the findings.

No definite guidelines have been established for dividing the sample into analysis and holdout groups. The most popular procedure is to divide the total group so that one-half of the respondents are placed in the analysis sample and the remaining half are placed in the holdout sample. However, no hard and fast rule has been established here and some researchers prefer a 60–40 or a 75–25 split between the analysis and holdout groups.

When selecting the individuals for the analysis and holdout groups, a proportionately stratified sampling procedure is usually followed. If the categorical groups for the discriminant analysis are equally represented in the total sample, then an equal number of individuals is selected. If the categorical groups are unequal, then the sizes of the groups selected for the hold-out sample should be proportionate to the total sample distribution. If a sample consists of 50 males and 50 females, then the hold-out sample would have 25 males and 25 females. If the sample contained 70 females and 30 males, then the hold-out samples would consist of 35 females and 15 males.

Several additional comments need to be made regarding the division of the total sample into analysis and holdout groups. One is that if the analyst is going to divide the sample into analysis and hold-out groups, then the sample must be sufficiently large to do so. Again no hard and fast rules have been established, but it seems logical the analyst would want at least a hundred in the total sample to justify dividing it into the two groups. One compromise procedure the analyst can select if the sample size is too small to justify a division into analysis and holdout groups is to develop the function on the entire sample and then use the function to classify the same group used to develop the function. Such a procedure results in an upward bias in the predictive accuracy of the function, but is certainly better than not testing the function at all.

Recall that the most frequent procedure utilized in validating the discriminant function is to divide the groups randomly into analysis and holdout samples a single time. Such a procedure would involve developing a discriminant function with the analysis sample and then applying it to the holdout sample. Other researchers have suggested, however, that greater confidence could be placed in the validity of the function by following this procedure several times [5]. Instead of randomly dividing the total sample into analysis and holdout groups one time, the analyst would randomly divide the total sample into analysis and holdout samples several times, each time testing the validity of the function through the development of a classification matrix and hit ratio. Then the several hit ratios would be averaged to obtain a single measure.

Other more sophisticated methods have been suggested for validating discriminant functions. The selection of a validation method depends upon the amount of time and facilities available and the de-gree to which the researcher wishes to follow a rigorous analysis procedure. A summary of more rigorous methods may be found in Crask and Perreault [3].

Computational Method. Two computational methods can be utilized in deriving discriminant function, the simultaneous (direct) method and the stepwise method.

The *simultaneous method* involves computing the discriminant function so that all the independent variables are considered concur-rently. Thus, the discriminant function(s) is computed based upon the entire set of independent variables, regardless of the discriminating power of each of the independent variables. The simultaneous method is appropriate when, for theoretical reasons, the analyst wants to in-clude all the independent variables in the analysis and is not interested in seeing intermediate results based only on the most discriminating variables [10].

The *stepwise method* is an alternative to the simultaneous approach, and involves entering the independent variables into the discriminant function one at a time on the basis of their discriminating power. The stepwise approach begins by choosing the single best discriminating variable. The initial variable is then paired with each of the other independent variables one at a time, and a second variable is chosen. The second variable is the one which is best able to improve the discriminating power of the function in combination with the first variable. The third and any subsequent variables are selected in a similar manner. As additional variables are included, some previously selected variables may be removed if the information they contain about group differences is available in some combination of the other included variables. Eventually, either all independent variables will have been included in the function or the excluded variables will have been judged as not contributing significantly toward further discrimination [10].

The stepwise method is useful when the analyst wants to consider a relatively large number of independent variables for inclusion in the function. By sequentially selecting the next best discriminating variable at each step, variables which are not useful in discriminating between the groups are eliminated and a reduced set of variables is identified. The reduced set typically is almost as good as, and sometimes better than, the complete set of variables.

Statistical Significance.   After the discriminant function has been computed, the analyst must assess its level of significance. The conventional criterion of .05 or beyond is used. If the function is not significant at or beyond the .05 level, then there is little justification for going further. This is because there is little likelihood that the function will classify more accurately (that is, with fewer misclassifications) than would be expected by randomly classifying individuals into groups.

Stage Two: Validation

The validation stage involves several major considerations. These considerations are: why develop classification matrices, cutting score determination, constructing classification matrices, chance models, and classification accuracy relative to chance.

Why develop classification matrices.   As has been noted, one of the standard outputs of a discriminant analysis is a measure of the statistical significance of the function. For the Veldman [12] and SPSS [10] packages the statistic is a Chi-Square. For the BMD package [4] it is the Mahalanobis D^2 statistic. Although these statistics determine the significance of the discriminant function, in reality they are weak tests and mean very little. For example, suppose the two groups are significantly different beyond the .01 level. With sufficiently large sample sizes, the group means (centroids) could be virtually identical and we still would have statistical significance. In short, these statistics suffer the same drawbacks of classical tests of hypotheses. Thus, the level of significance of these statistics is a very poor indication of the function's ability to discriminate between the two groups.

To further clarify the usefulness of the classification matrix procedure, we shall relate it to the concept of an R^2 in regression analysis. Most of us have probably read academic articles in which the author has found statistically significant relationships, and yet has explained only 10 percent (or less) of the variance; i.e., $R^2 = 0.10$. Usually this R^2 is significantly different from zero simply because the sample size is large. With multiple discriminant analysis, the hit-ratio (percentage correctly classified) is analogous to regression's R^2. The hit-ratio reveals how well the discriminant function classified the statistical units; the R^2 indicates how much variance the regression equation explained. The F-Test for statistical significance of the R^2 is, therefore, analogous to the Chi-Square (or D^2) test of significance in discriminant analysis. Clearly, with a sufficiently large sample size in discriminant analysis, we could have a statistically significant difference between the two (or more) groups, and yet correctly classify only 53 percent (when chance is 50 percent, with equal group sizes) [9].

Cutting score determination. If the statistical test indicates that the function discriminates significantly, then it is customary to develop classification matrices to provide a more accurate assessment of the discriminating power of the function. Before a classification matrix can be constructed, however, the analyst must determine the cutting score. The *cutting score* is the criterion (score) against which each individual's discriminant score is judged to determine into which group the individual should be classified.

In constructing classification matrices, the analyst will want to determine the *optimum cutting score* (also referred to as a critical Z value). The optimal cutting score will differ depending on whether the sizes of the groups are equal or unequal. If the groups are of equal size, then the optimal cutting score will be half-way between the two group centroids. The cutting score is therefore defined as:

$$Z_{CE} = \frac{\overline{Z}_A + \overline{Z}_B}{2}$$

where

Z_{CE} = critical cutting score value for equal group sizes
$\overline{Z}_A$ = centroid for group A
$\overline{Z}_B$ = centroid for group B

If the groups are not of equal size, then a weighted average of the group centroids will provide an optimal cutting score, calculated as follows:

$$Z_{CU} = \frac{N_B\overline{Z}_A + N_A\overline{Z}_B}{N_A + N_B}$$

where

Z_{CU} = critical cutting score value for unequal group sizes
N_A = number in group A
N_B = number in group B
$\overline{Z}_A$ = centroid for group A
$\overline{Z}_B$ = centroid for group B

Both of the above formulas assume the distributions are normally distributed and the group dispersion structures are known.

The concept of an optimal cutting score is illustrated in Figures 3.3 and 3.4. The optimal cutting score for equal groups is shown in Figure 3.3. The effect of one group being larger than the other is illustrated in Figure 3.4. Both the weighted and the unweighted cutting scores are shown. It is apparent that if the size of group A is much smaller than group B, then the optimal cutting score will be closer to the centroid of group A than it is to the group B centroid. Also, note that if the unweighted cutting score were used, none of the individuals in group A would be misclassified, but a substantial portion of those in group B would.

The optimal cutting score also must consider the costs of misclassifying an individual into the wrong group. If the costs of misclassifying an individual are approximately equal, then the optimal cutting score will be the one which will misclassify the fewest number of individuals in all groups. If the misclassification costs are unequal, then the optimum cutting score will be the one which minimizes the costs of misclassification.

More sophisticated approaches to determining cutting scores are discussed in Massy [8]. The approaches are based upon a Bayesian

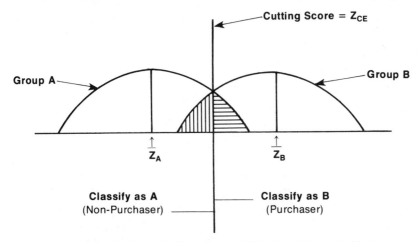

Figure 3.3—Optimal Cutting Score With Equal Sample Sizes

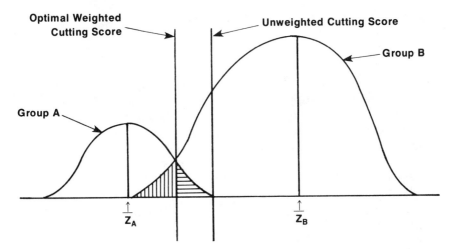

Figure 3.4—Optimal Cutting Score With Unequal Sample Sizes

statistical model and are appropriate when the costs of misclassification into certain groups are very high, when the groups are of grossly different sizes, or when one wants to take advantage of a priori knowledge of group membership probabilities.

In practice, when calculating the cutting score, it is usually not necessary to insert the raw variable measurements for every individual into the discriminant function and obtain the discriminant score for each person to use in computing the $\bar{Z}_A$ and $\bar{Z}_B$ (group A and B centroids). In many instances the computer program will provide the discriminant scores as well as the $\bar{Z}_A$ and $\bar{Z}_B$ as regular output. When the analyst has the group centroids and sample sizes, he must merely substitute the values in the appropriate formula to obtain the optimal cutting score.

Constructing classification matrices.   To validate the discriminant function through the use of classification matrices, the sample should be randomly divided into two groups. One of the groups (the analysis sample) is used to compute the discriminant function. The other group (the holdout or validation sample) is retained for use in developing the classification matrix. The procedure involves multiplying the weights generated by the analysis sample times the raw variable measurements of the hold-out sample to obtain discriminant scores for the hold-out sample. Then the individual discriminant scores for the hold-out sample are compared with the critical cutting score value and classified as follows:

(1) Classify an individual into group A if $Z_n < Z_{ct}$.
(2) Classify an individual into group B if $Z_n > Z_{ct}$.
where

Z_n = the discriminant Z-score for the nth individual
Z_{ct} = the critical cutting score value

The results of the classification procedure are presented in matrix form as shown in Table 3.2. The entries on the diagonal of the matrix represent the number of individuals correctly assigned to their group. The numbers off the diagonal represent the incorrect classifications. The entries under the column labeled Actual Total represent the number of individuals actually in each of the two groups. The entries at the bottom of the columns represent the number of individuals as-

TABLE 3.2
CLASSIFICATION MATRIX FOR HATCO'S
NEW CONSUMER PRODUCT

| ACTUAL GROUP | PREDICTED GROUP | | Actual Total | Group Classification Percentage |
	Would Purchase (1)	Would Not Purchase (2)		
(1)	22	3	25	88
(2)	5	20	25	80
Predicted Total	27	23	50	

Percent Correctly Classified (Hit-Ratio) $=$ (100) (22 $+$ 20/50) $=$ 84%

signed to the two groups by the discriminant function. The percentage correctly classified for each group is shown at the right side of the matrix and the overall percentage correctly classified (Hit-Ratio) is shown at the bottom of the matrix. For example, in our HATCO problem which attempted to predict which housewives would purchase a new consumer product, the number of individuals actually in and correctly assigned to Actual Group (1) Would Purchase—was 22. The number incorrectly assigned to Actual Group (2) Would Not Purchase—was 3. Similarly, the number of correct classifications to Group (2) was 20 and the number of incorrect assignments to group (1) was 5. Thus, the percentage classification accuracy of the discriminant function for Groups (1) and (2) would be 88 and 80 percent, respectively. The over-all classification accuracy (hit ratio) would be 84 percent.

One final classification procedures topic should be discussed. A t-test is available to determine the level of significance for the classification accuracy. The formula for a two group analysis (equal sample size) is:

$$t = \frac{p - .5}{\sqrt{\dfrac{.5(1 - .5)}{N}}}$$

where

p = proportion correctly classified
N = sample size

This formula can be adapted for use with more groups and unequal sample sizes. For details of these adaptations see Anderson [1].

Chance models. This section on chance models describes the procedure involved in determining the percent of individuals that would be correctly classified by chance. As noted earlier, the predictive accuracy of the discriminant function is measured by the hit-ratio, which is obtained from the classification matrix. The analyst may ask what is considered an acceptable level of predictive accuracy for a discriminant function, and what is not acceptable? For example, is 60 percent an acceptable level, or should one expect to obtain 80 to 90 percent predictive accuracy? To answer this question the analyst must first determine the percentage that could be classified correctly by chance (without the aid of the discriminant function).

When the sample sizes of the groups are equal, the determination of the chance classification is rather simple, obtained by dividing 1 by the number of groups. The formula is: $C = 1/$ number of groups. For instance, in a two group function the chance probability would be .50; for a three group function the chance probability would be .33, and so forth.

The determination of the chance classification for situations where the group sizes are unequal is somewhat more involved. Let's assume we have a sample in which 75 subjects belong to one group and 25 to the other. We could arbitrarily assign all the subjects to the larger group and achieve a 75 percent classification accuracy—without the aid of a discriminant function. It could be concluded that unless the discriminant function achieves a classification accuracy higher than 75 percent, it should be disregarded because it has not helped us improve our prediction accuracy.

Determining the chance classification based on the sample size of the largest group is referred to as the *maximum chance criterion*. It is determined by computing the percentage of the total sample represented by the largest of the two (or more) groups. For example, if the group sizes are 65 and 35, then the maximum chance criterion is 65 percent correct classifications. Therefore, if the hit ratio for the discriminant function does not exceed 65 percent, it has not helped us predict—based on this criterion.

The maximum chance criterion should be used when the sole objective of the discriminant analysis is to maximize the percentage correctly classified [9]. Situations for which we are concerned only about maximizing the percentage correctly classified are rare. Usually the analyst uses discriminant analysis to correctly identify members of both groups. In cases where the sample sizes are unequal and the analyst wants to classify members of both groups, the discriminant function defies the odds by classifying a subject in the smaller group. The chance criterion should take this into account [9]. Therefore, another chance model—the *proportional chance criterion*—should be used in most situations.

The proportional chance criterion should be used when group sizes are unequal, and the analyst wishes to correctly identify members of the two (or more) groups. The formula for this criteria is:

$$C \text{ proportional} = p^2 + (1 - p)^2$$

where

$\quad\quad p =$ the proportion of individuals in group 1
$1-p =$ the proportion of individuals in group 2.

Using the group sizes from our earlier example (75 and 25) the proportional chance criterion would be 62.5 percent compared to 75 percent. Therefore, a prediction accuracy of 75 percent might be acceptable because it is above the 62.5 percent proportional chance criterion.

It should be noted that these chance model criteria are useful only when computed with hold-out samples (split sample approach). If the individuals used in calculating the discriminant function are the ones being classified, the result will be an upward bias in the prediction accuracy [5]. In such cases both of these criteria would have to be adjusted upward to account for this bias.

Classification Accuracy Relative to Chance. The question of classification accuracy is crucial. If the percentage of correct classifications is significantly larger than would be expected by chance, an attempt can be made to interpret the discriminant functions in the hope of developing group profiles. However, if the classification accuracy is no greater than can be expected by chance, whatever structural differences appear to exist merit little or no interpretation [5]. That is, differences in score profiles would provide no meaningful information for identifying group membership.

The question then is how high should the classification accuracy be relative to chance? For example, if chance is 50 percent (two group, equal sample sizes), does a classification (predictive) accuracy of 60 percent justify moving to the interpretation stage? No general guidelines have been developed to answer this question. Ultimately the decision rests on the cost versus the value of the information. If the costs associated with a 60 percent predictive accuracy (relative to 50 percent by chance) are greater than the value to be derived from the findings, then there is no justification for interpretation. If the value is high relative to the costs, then 60 percent accuracy would justify moving on to interpretation.

The cost versus value argument offers little assistance to the neophyte data analyst. Therefore, the authors suggest the following criterion: the classification accuracy should be at least 25 percent greater than by chance. For example, if chance accuracy is 50 percent the classification accuracy should be 62.5 percent. If chance accuracy is 30 percent, then the classification accuracy should be 37.5 percent. This criterion provides only a rough estimate of the acceptable level of predictive accuracy. The criterion is easy to apply with groups of equal size. With unequal size groups an upper limit is reached when the maximum chance model is used to determine chance accuracy. This does not present too great a problem, however, since under most circumstances the maximum chance model would not be used with unequal group sizes (see section on chance models).

Stage Three: Interpretation

If the discriminant function is statistically significant and the classification accuracy is acceptable, then the analyst should continue to Stage Three which focuses on making substantive interpretations of the findings. This involves examining the discriminant functions to determine the relative importance of each of the independent variables in discriminating between the groups. Three methods have been proposed: (1) standardized discriminant weights, (2) discriminant structure correlations, and (3) partial F-values.

Discriminant weights. The traditional approach to interpreting discriminant functions involves examining the sign and magnitude of the standardized discriminant weight (sometimes referred to as a discriminant coefficient) assigned to each variable in computing the discriminant functions. Independent variables with relatively larger weights contribute more to the discriminating power of the function than do variables with smaller weights. Therefore, when the sign is ignored, each weight represents the relative contribution of its associated variable to that function. The sign merely denotes that the variable makes either a positive or negative contribution [10].

The interpretation of discriminant weights is analogous to the interpretation of beta weights in regression analysis, and is therefore subject to the same criticisms. For example, a small weight may either mean its corresponding variable is irrelevant in determining a relationship, or that it has been partialed out of the relationship because of a high degree of multicollinearity. Another problem with the use of discriminant weights is that they are subject to considerable instability. These problems suggest caution in using weights to interpret the results of discriminant analysis.

Discriminant Loadings. In recent years loadings have increasingly been used as a basis for interpretation because of the deficiencies in utilizing weights. Discriminant loadings, referred to sometimes as structure correlations, measure the simple linear correlation between each independent variable and the discriminant function. The discriminant loadings reflect the variance the independent variables share with the discriminant function, and can be interpreted like factor loadings in assessing the relative contribution of each independent variable to the discriminant function (Chapter 6 further discusses factor loadings interpretation).

Discriminant loadings (like weights) may be subject to instability. Loadings are considered relatively more valid than weights as a means of interpreting the discriminating power of independent variables.

The analyst still must be cautious when using loadings for interpreting discriminant functions.

Partial F-values. As discussed earlier, two computational approaches can be utilized in deriving discriminant functions—simultaneous and stepwise. When the stepwise method is selected, an additional means of interpreting the relative discriminating power of the independent variables is available through the use of the partial F-values. This is accomplished by examining the absolute sizes of the significant F-values and ranking them. Larger F-values would indicate greater discriminating power.

In practice, rankings using the F-values approach would be the same as with the weights, but the F-values would indicate the associated level of significance for each variable.

Which interpretation method to use. Several methods for interpreting the nature of discriminant functions have been discussed. Which method should the analyst use? Since most discriminant analysis problems necessitate use of a computer, the analyst frequently must use whichever method is available on the system packages. The widely available SPSS [10] and BMD [4] packages provide the discriminant coefficients and the F-values with the stepwise options. The less well known Veldman [12] and Cooley and Lohnes [2] packages provide both the weights and the discriminant loadings. The loadings approach is somewhat more valid than the use of weights, and should be utilized whenever possible. The analyst should not hesitate to use the weights if they are the only method available, but they should be interpreted in light of their limitations.

AN ILLUSTRATIVE EXAMPLE

To illustrate the application of discriminant analysis we shall use variables drawn from the data bank introduced in Chapter 1. A HATCO policy is to present an award to the top salesperson each month to motivate them to sell more. HATCO would like to predict which individuals are likely to receive this award. This information would assist HATCO in hiring new salespersons. HATCO believes it can use the battery of social-psychological tests administered to all new sales recruits to identify potential recipients of the top salesperson award. The problem involves using the first six variables from the test bank to predict variable X_{10}, a categorical variable which indicates which salespeople have received the top salesperson award in the past. By developing a discriminant function using existing salespersons records,

HATCO hopes to identify new sales recruits bearing a high success potential.

Previous discussion has emphasized the need for validating a discriminant function through the use of a split-sample approach. Our data bank consists of only 50 observations. Therefore, in this example a split sample approach was not utilized. A simultaneous discriminant analysis was performed with the HATCO data (all variables entered concurrently). The metric independent variables were the six social-psychological measures (variables $X_1 - X_6$) and the categorical dependent variable was X_{10}—top salesperson award. HATCO records indicated that 30 persons had received the award and 20 had not. Thus, the dependent variable consisted of two groups—one with a sample size of 30 and the other with 20.

Let's begin our analysis of the HATCO discriminant analysis problem by examining Table 3.3. The data are presented to enable a comparison between the profiles for the two groups. At the top are the centroids (weighted group means for all variables) and below are the unweighted group means for each separate independent variable. To the right are the tests for significant differences between the group means.

Recall that the last step in the derivation stage of a discriminant analysis is assessing the level of significance of the function. The dis-

TABLE 3.3
GROUP MEANS AND SIGNIFICANCE TESTS
FOR HATCO DISCRIMINANT ANALYSIS

| | Group Means | | Significance Tests |
	Non-Recipients	Recipients	
Centroids	10.27	14.54	40.42[1]
Independent Variables			
X_1 Self-Esteem	4.75	8.30	32.23[2]
X_2 Locus of Control	5.79	3.82	9.95
X_3 Alienation	13.40	17.17	37.17
X_4 Social			
Responsibility	10.40	10.36	0.01
X_5 Machiavellianism	5.26	6.02	3.56
X_6 Political Opinion	5.07	5.31	0.30

[1] The test for the equality of group centroids is a generalized Chi-square. It is significant beyond the 0.0000 level which indicates the groups differ significantly.

[2] The tests for the independent variables are univariate F-ratios. Variables X_1, X_2, and X_3 are significant beyond the 0.003 level; the other variables are not significantly different.

criminant function is utilized to test the hypothesis that the group centroids are equal (not significantly different). Our findings reveal that the group centroids for our HATCO problem are significantly different. A Chi-square value of 40.42 (with 6 degrees of freedom) is highly significant—beyond the 0.000 level. The sizes of the centroids indicate that those individuals who have received the Top Salesperson Award generally have higher scores on the social-psychological measures. Recipients have a centroid of 14.54 whereas nonrecipients have a centroid of 10.27. Specific interpretation of the sizes of the group centroids is possible. However, the analyst must be cautious in doing this because the meaning of different sizes is dependent on the method of scoring for the individual independent variables. Discussion of the interpretation of centroid sizes will be covered in the interpretation stage after the function has been validated. Since the function is statistically significant, it is logical to move to the validation stage of discriminant analysis.

The second stage in a discriminant analysis is validation of the function. To accomplish this, we would develop a classification matrix to assess the predictive accuracy of the function. Before the classification matrix can be constructed, we must determine the appropriate cutting score. Recall that the cutting score is the criterion against which each individual's discriminant score is judged to determine into which group the individual should be classified.

The group of recipients for the Top Salesperson Award contains 30 individuals and the non-recipient group contains 20 persons. Therefore, the cutting score formula for unequal size groups must be used:

$$Z_{CU} = \frac{N_B \bar{Z}_A + N_A \bar{Z}_B}{N_A + N_B}$$

where

Z_{CU} = critical cutting score value for unequal group sizes
N_A = number in group A
N_B = number in group B
$\bar{Z}_A$ = centroid for group A
$\bar{Z}_B$ = centroid for group B

By substituting the appropriate numbers in the formula we can obtain the critical cutting score (assuming equal costs of misclassification):

$$Z_{CU} = \frac{(30)\,(10.27) \;+\; (20)\,(14.54)}{20 \;+\; 30}$$

$$Z_{CU} = \frac{308.1 \;+\; 290.8}{50} = \frac{589.9}{50} = 11.978$$

The critical cutting score is 11.978. Our procedure for classifying individuals will be:

(1) Classify an individual as a non-recipient if their discriminant score is less than 11.978.

TABLE 3.4
DISCRIMINANT SCORES FOR HATCO SALES FORCE

Number	Non Recipients	Recipients
(1)	10.72	10.18*
(2)	12.14*	11.07*
(3)	10.63	12.25
(4)	10.44	13.09
(5)	10.01	9.68*
(6)	10.40	14.85
(7)	9.32	13.45
(8)	8.14	15.76
(9)	7.85	18.60
(10)	10.84	15.25
(11)	9.97	14.62
(12)	10.99	17.13
(13)	11.08	14.87
(14)	8.37	15.80
(15)	11.27	17.65
(16)	12.14*	13.52
(17)	10.02	12.32
(18)	11.48	13.52
(19)	11.33	16.81
(20)	9.29	13.31
(21)		15.10
(22)		16.01
(23)		14.84
(24)		14.79
(25)		16.59
(26)		16.33
(27)		15.35
(28)		12.97
(29)		14.79
(30)		15.63

*asterisks indicate individuals misclassified.

(2) Classify an individual as a recipient if their discriminant score is greater than 11.978.

Using the critical cutting score and the individual discriminant scores provided in Table 3.4, we can proceed to develop a classification matrix. Two persons in the non-recipient group are above the critical score, and three persons in the recipient group are below. Thus, the discriminant function classified 18 of 20 non-recipients correctly and 27 of 30 recipients correctly—for a classification accuracy of 90 percent. These results are displayed in matrix form in Table 3.5. The number of correct classifications is shown on the diagonal and the number of incorrect ones off the diagonal. Individual group classifications are shown on the right and at the bottom is the hit ratio. In this example both group accuracies and overall hit-ratio are the same. In most instances the groups will differ, and this hit-ratio will represent an average.

The 90 percent classification accuracy is quite high, but for illustrative purposes we shall compare it with the a priori chance of classifying individuals correctly without our discriminant function. The proportional chance criterion is the appropriate chance model to use for our HATCO example. We have unequal group sizes and we want to correctly identify members of both groups. The formula is:

$$C_{pro} = p^2 + (1-p)^2$$

where

$$
\begin{aligned}
C_{pro} &= \text{the proportional chance criterion} \\
p &= \text{proportion of individuals in group 1} \\
1-p &= \text{proportion of individuals in group 2}
\end{aligned}
$$

TABLE 3.5

CLASSIFICATION MATRIX FOR HATCO SALES FORCE

| ACTUAL GROUP | Predicted Group | | Actual Total | Group Classification Percentage |
	Non-Recipients (1)	Recipients (2)		
(1)	18	2	20	90
(2)	3	27	30	90
Predicted Total	21	29	50	

Percent Correctly Classified (Hit-ratio) = (100) (18 + 27/50) = 90%

Substituting the appropriate numbers in the formula we obtain.

$$C_{pro} = (.40)^2 + (.60)^2$$
$$C_{pro} = 16 + 36$$
$$C_{pro} = 52 \text{ percent}$$

The classification accuracy of 90 percent is substantially higher than the proportional chance criteria of 52 percent. The discriminant function can be considered as a valid predictor of recipients versus non-recipients. However, the analyst still should not be over-optimistic about the 90 percent because of the upward bias from classifying the same individuals as used in computing the function.

After validating the function the next phase is interpretation. The interpretation phase involves examining the function to determine the relative importance of each of the independent variables in discriminating between the groups. Table 3.6 contains the discriminant weights and loadings for the function. The first step in interpretation is to rank the independent variables in terms of their relative discriminatory power. The rankings are based on the absolute sizes of either the loadings or the weights. Signs do not affect the rankings; they indicate a positive or negative relationship with the dependent variable. Since the loadings are considered somewhat more valid than weights we shall use them in our example. The weights are displayed for comparison purposes. A comparison of the two approaches reveals substantial differences in rankings, particularly for variables X_2 and X_6. Indeed, the rankings are the same for only one variable—X_5.

TABLE 3.6
HATCO DISCRIMINANT ANALYSIS RELATING
SOCIAL-PSYCHOLOGICAL VARIABLES TO TOP SALESPERSON AWARD

Independent Variables	Discriminant Weights	Rank	Discriminant Loadings	Rank
X_1 Self-Esteem	0.712	1	0.858	2
X_2 Locus of Control	0.080	6	−0.542	3
X_3 Alienation	0.543	2	0.864	1
X_4 Social Responsibility	−0.081	5	−0.013	6
X_5 Machiavellianism	−0.300	4	0.343	4
X_6 Political Opinion	0.308	3	0.104	5

When using the discriminant loadings interpretation approach, we need to know which variables are significant discriminators. Generally any variables exhibiting loadings of $\pm$ .30 or higher are considered significant. Other considerations are possible, however (see Chapter 6: Factor Analysis). Using this simplified criteria, four variables—X_1, X_2, X_3, and X_5—are significant. Some analysts would not consider variable X_5 significant because of the substantial difference between it and variable X_2 (.34 versus $-$.54).

The analyst usually is interested in substantive interpretations of the individual variables. This is accomplished by identifying the variables and understanding how they are scored (i.e., what high scores and low scores mean). The four significant variables and their scoring procedures are: X_1 Self-Esteem—high score = high self-esteem; X_2 Locus of Control—high score = external locus; X_3 Alienation—high scores = not alienated, and X_5 Machiavellianism—high score = highly Machiavellian. For more detail on the meaning of these variables see Chapter 1. Reviewing Table 3.3 we note that the "Recipients" score higher (large mean) relative to the "Non-Recipients" on Self-Esteem, Alienation, and Machiavellianism. They score lower (smaller mean) on Locus of Control. Using this information we could substantively interpret these findings to profile the characteristics of Recipients and Non-Recipients. Specifically, our findings indicate that Recipients exhibit higher self-esteem, an internal locus of control, are not alienated, and are somewhat more Machiavellian than are Non-Recipients. With this information, HATCO could select new salespersons that fit the Recipients profile and expect to be more successful in hiring good salespeople.

Summary

The underlying nature, concepts and approach have been presented for multiple discriminant analysis. Basic guidelines for application and interpretation were included to further clarify the methodological concepts. An illustrative example based on the first chapter data demonstrated the major points you need to be familiar with in applying discriminant analysis.

Multiple discriminant analysis helps you to understand and explain research problems which involve a single categorical dependent variable and several metric independent variables. A mixed data set (both metric and non-metric) is also possible for the independent variables if the non-metric variables are dummy coded (zero-one). The result of a discriminant analysis can assist you in profiling the intergroup characteristics of the subjects and in assigning them to their appropriate groups. Potential applications of discriminant analysis to both business and non-business problems are numerous.

END OF CHAPTER QUESTIONS

1. How would you differentiate between the statistical techniques of multiple discriminant analysis, regression analysis, and analysis of variance?
2. What criteria could you use in deciding whether or not to stop a discriminant analysis after the derivation stage? After the validation stage?
3. What procedure would you follow in dividing your sample into analysis and holdout groups. How would you change this procedure if your sample consisted of fewer than 100 individuals or objects.
4. How would you go about determining the optimum cutting score?
5. How would you determine whether or not the classification accuracy of the discriminant function is sufficiently high relative to chance classification?

REFERENCES

1. Anderson, T. W., *Introduction to Multivariate Statistical Analysis,* New York: John Wiley & Sons, Inc., 1958.
2. Cooley, W., Cooley, W. W. and P. R. Lohnes, *Multivariate Data Analysis.* New York: John Wiley and Sons, Inc., 1971.
3. Crask, M. and W. Perreault, "Validation of Discriminant Analysis in Marketing Research," *Journal of Marketing Research,* Vol. XIV, February, 1977.
4. Dixon, W. J., *Biomedical Computer Programs.* Los Angeles: University of California Press, 1967.
5. Frank, R. E., W. F. Massey and D. G. Morrison, "Bias in Multiple Discriminant Analysis," *Journal of Marketing Research,* Vol. 2, No. 3, August, 1965, pp. 250–58.
6. Green, Paul E., and Donald S. Tull, *Research for Marketing Decisions.* Englewood Cliffs: Prentice-Hall, Inc., 1975.
7. Harris, R. J., *A Primer of Multivariate Statistics,* New York: Academic Press, 1975.
8. Massey, W. F., "Bayesian Multiple Discriminant Analysis," Graduate School of Business, Stanford University, Working Paper No. 58, July, 1965.
9. Morrison, Donald G., "On the Interpretation of Discriminant Analysis," *Journal of Marketing Research,* Vol. 6, No. 2, May 1969, pp. 156–63.
10. Nie, N., C. Hull, D. Bent, M. Nieswonger, *Statistical Package for the Social Sciences,* New York: McGraw-Hill, 1975.

11.   Overall, J. E., and J. Klett, *Applied Multivariate Analysis*. New York: McGraw-Hill, Inc., 1972.
12.   Veldman, D., *Fortran Programming for the Behavioral Sciences*. New York: Holt, Rinehart, Winston, Inc., 1967.

SELECTED READINGS

BANK CREDIT CARD USERS: NEW MARKET SEGMENT FOR REGIONAL RETAILERS

JAC L. GOLDSTUCKER and ELIZABETH C. HIRSCHMAN

Over the past few years much research has been directed toward developing ways of identifying market segments. The purpose of many of these segmentation studies has been to establish criteria for marketing strategies to reach target markets more efficiently and more effectively.

A market segment of interest in recent years has been that of bank credit card users. Initially, credit cards were mailed almost indiscriminantly to bank customers with little regard to identifying the characteristics of those people most likely to comprise an appropriate target market. The chaos which resulted provided an incentive for banks to reassess their policy. In addition, it suggested to marketing scholars an area in which research might help refine marketing strategies to identify yet another and probably important segment.[1]

Recently, the bank card user segment has become particularly important to regional retailers. This importance results from the mobility of the U.S. population and the relationship of the mobile sector to regional retail department and specialty store chains. Proprietary research conducted by a major department store chain indicates, first, that many newcomers to a market area not only have charge cards issued by national chain department stores such as J. C. Penney and Sears, Roebuck and Company but also have bank credit cards.[2] Newcomers tend to make initial purchases at a local branch of the national chain store with which they already have established credit. Because regional and local stores lack this national base of credit card customers, they are at a disadvantage in attracting newcomers.

Second, the research indicates that the inflow of tourists and con-

Jac L. Goldstucker is on the faculty in marketing at Georgia State University. Elizabeth C. Hirschman is a graduate student in marketing there.

Jac L. Goldstucker and Elizabeth C. Hirschman, "Bank Credit Card Users: A New Market Segment for Regional Retailers," pp. 5–11, *MSU Business Topics*, Summer 1977. *Reprinted by permission of the publisher,* Division of Research, Graduate School of Business Administration, Michigan State University.

vention and business visitors is an important source of revenue to many communities. While away from home, these groups patronize the host city's shops and stores. They rely on such credit cards as Bank-Americard, Master Charge, and American Express since they seldom have credit cards from the local retailers. Therefore, these people gravitate toward those local stores that accept bank credit cards, again putting the regional retailers at a disadvantage.

Third, there is a segment of the market which prefers not to be overburdened with credit cards. Persons in this segment prefer to carry one or two multipurpose cards. Often, a bank card is one of those carried.

Purpose of the study

For the reasons just enumerated, and perhaps for others as well, in 1975 several major regional department store chains, led by Rich's of Atlanta, began to accept bank charge cards in addition to their own cards.[3] The strategy apparently has been successful and appears likely to continue. By late 1976, 34 stores among the 100 top volume U.S. department stores had begun accepting bank charge cards. This strategy has opened a new market for bank credit card services.

It is probable that greater insight into the characteristics of bank card users will enable regional retailers to devise marketing strategies directed at capturing a larger share of that market segment and thus to compete more effectively with national retail chain stores.

The purpose of this study is to describe bank card users in terms of the store image characteristics they seek and the social activities in which they participate. This information can provide executives of regional department stores, as well as other regional and local retailers, with insights into the kind of retail store bank credit card users find most appropriate to their shopping needs. Armed with the information, these regional department stores will be able to define this market segment more precisely and develop market strategies to attract it.

Methodology

Demographic data have been widely used in developing profiles of bank credit card users. Some researchers have used psychographic variables as well.[4] The variables of store image characteristics and social activity have not been used to study these card users. Nonetheless, it is reasonable to assume that these two groups of variables can provide insight into the shopping behavior of bank card users and their expectations with respect to store offerings.

Knowledge of the store characteristics a bank card holder uses in deciding where to shop appears to be relevant to the retailers seeking to attract this market segment. Furthermore, knowledge of the social activities in which card users participate can give retailers greater insight into life-styles of their customers and thus help them formulate marketing strategies.

The sample.   A major regional department store chain seeking to establish base-line data for several merchandise lines and store image perceptions conducted a large-scale survey in its home market of Atlanta, Georgia. From a total sample of 1,200 in the Atlanta Standard Metropolitan Statistical Area (SMSA), 952 usable telephone interviews were obtained. A random digit dialing technique eliminated the sampling bias against new arrivals and unlisted numbers associated with use of the telephone directory. The sample was taken according to population distribution. To ensure adequate representation of working persons, all interviews were conducted between 6 P.M. and 9:30 P.M. on weeknights.

The independent variables.   Three sets of variables were selected from the data for use as predictors to profile bank credit card users. These sets included store image characteristics and social activity variables as well as standard demographic variables. They are listed in Table 1. The measurement scales used for each variable also are provided in Table 1.

The dependent variable.   The dichotomous criterion of bank card user/nonuser was determined by responses to the question: "Do you have a BankAmericard, Master Charge, or C & S (local bank card) charge card? " Respondents who answered yes with respect to one or all of the three cards were classified as bank card users. There were 455 (47.8 percent) bank card users and 497 (52.2 percent) bank card nonusers in the sample.

The analysis

Step-wise discriminant analysis was used to determine those demographic characteristics, store attribute evaluations, and social activities which typify bank card users. Since discriminant analysis provides a means of distinguishing statistically between two or more groups, it is a useful technique in developing customer profiles. To distinguish among several customer groups or between a particular customer group and the general population, the researcher selects a collection of descriptive variables that measure characteristics on which the groups are expected to differ. The mathematical objective of discriminant analysis is to weight and linearly combine these descriptive vari-

TABLE 1
SEGMENTATION VARIABLES

Demographic	Sociographics[a]	Store image[b]
Possession of national department store credit cards[c]	Gardening	Sales clerk service
	Hunting/fishing	Store location
	Camping/back-packing	Merchandise pricing
Possession of regional department store credit cards[c]	Golf	Credit or billing policies
	Tennis	Store layout or atmo-sphere
Possession of discount store credit cards[c]	Water skiing/swimming	Merchandise quality
BankAmericard[c]	Entertain in home	Merchandise variety and assortment
Master Charge[c]	Bridge clubs (member)	Merchandise display
Local bank credit card[c]	Photography	Store's guarantee, ex-change, or adjust-ment policies
Possession of no credit card[c]	Concerts/ballet (attends)	Store sales represent real savings
Marital status (married/not)[c]	Rock concerts (attends)	
Number of children liv-ing at home[d]	Plays (attends)	
Children in various age groups[c]	Movies (attends)	
Spouse's age[d]	Book clubs	
Respondent's age[d]	Spectator sporting events (attends)	
House as a residence[c]	Membership in social organizations	
Length of time lived in metropolitan area[c]	Membership in reli-gious organizations	
Respondent's educa-tion (years)[d]	Membership in busi-ness organizations	
Spouse's education (years)[d]		
Income (household total)[d]		
Sex[c]		
Race (white/nonwhite)[c]		
Life cycle[e]		
Social class[f]		
Race (white/nonwhite)[c]		
Life cycle[e]		
Social class[f]		

[a] Measured on a dichotomous, 0/1 basis according to nonparticipation/participation.

[b] Assumed interval scale based on responses to question: "How important is (image - attribute) to you in deciding where to shop?" Very important, moderately important, not important. Responses were scored 3, 2, 1, respectively.

[c] Measured on a dichotomous, 0/1 basis according to possession or membership.

[d] Measured on an interval scale.

[e] Divided into six categories and coded dichotomously. The life cycle categories were: (1) single, under 34; (2) married, under 34, no children; (3) married, children under age 6; (4) married, all children over age 6; (5) married, over 34, no children at home; (6) single/widowed, over age 34.

[f] Social class is a weighted composite of income, education, and occupational status.

ables in some fashion so that the customer groups are differentiated as much as possible.[5] In this study, the customer groups consisted of bank card users and nonusers. The descriptive variables included demographic characteristics, social activities, and store image attribute evaluations.

Discriminant analysis provides two types of output that are especially valuable in profiling customer groups. First, it produces a discriminant function, or functions, representing a dimension along which the customer groups differ. The coefficients of the discriminating variables composing this function, when in standardized form, tell the relative importance of each of the variables.

The second valuable output results from the use of the discriminant function to classify individuals into various customer groups. In other words, once the discriminant function has been developed, it can be applied to a sample of persons, say, in a new market, and can predict how many will belong to a particular customer group.

These two abilities of discriminant analysis, the development of descriptive customer profile functions and the classification of individuals into customer groups, have obvious value to marketers. Once able to determine the major dimensions upon which customer groups differ, marketers can develop more effective strategies. Furthermore, when entering a new market, or in evaluating an existing one, marketers can determine what proportion of the population falls into each group.

The obvious utility of discriminant analysis in profiling customers has led to its widespread use and often abuse. An example of the potential dangers was presented more than ten years ago by R. E. Frank, W. F. Massey, and D. G. Morrison.[6] Using data on food product innovators as an example, the authors pointed out the existence of two potential sources of bias in discriminant analysis—sampling bias and search bias. The way to avoid these is simply to apply the obtained discriminant function to a new set of data. This was done here.

Findings

Table 2 reveals that the cross-validated discriminant functions correctly classified bank card users and nonusers at a level much beyond chance ($P < .0005$). A total of 77.9 percent of the bank card users and 77.4 percent of the nonusers have been correctly classified.

The standardized coefficients for the complex discriminant function profile are given in Table 3. As these coefficients indicate, the variables which discriminate most significantly between users and nonusers are (1) possession of regional department store charge cards,

TABLE 2
DISCRIMINANT ANALYSIS RESULTS

	Prior probability		Classification accuracy			
	Number	Percentage	Number	Percentage	T	P
Bank card nonuser	262	53.3	204	77.9	5.64	.0005
Bank card user	230	46.7	178	77.4	6.60	.0005

(2) possession of charge cards of J. C. Penney and of Sears, Roebuck and Company, (3) total family income, (4) possession of a discount store charge card, and (5) sex.

A further determination of the characteristics which differentiate bank card users and nonusers can be made by testing the differences between the group means for bank card users and nonusers on each independent variable. Using the F statistic, two of the store charac-

TABLE 3
STANDARDIZED DISCRIMINANT
FUNCTION COEFFICIENTS

Possession of regional department store charge card	.33
Possession of Sears charge card	.22
Possession of Penney charge card	.16
Total family income	.16
Possession of discount store charge card	.14
Spouse's education	.12
Entertain in home	.11
Gardening	.09
Membership in religious organization	.09
Play tennis regularly	.08
Importance of variety in stores' merchandise	.06
Live in house	.06
Attend plays	.05
Go camping/backpacking	−.05
Attend rock concerts	−.05
Have children under age 18 at home	−.06
Attend spectator sports	−.06
Photography as hobby	−.06
Race (black)	−.07
Sex (female)	−.14

NOTE: The standardized discriminant analysis function coefficients represent the relative importance of a particular variable in differentiating between bank card users and nonusers. Multicollinearity, or interrelatedness among the variables, can sometimes cause the coefficients to be unstable and potentially misleading. This was not felt to be a problem here, however, as correlations between the discriminant function variables were all below .40.

teristics and twelve of the social activity variables were found to differ significantly (p = .10) for the two groups. As indicated in Table 4, two significant store characteristics are variety in the store's merchandise and the store's credit or billing policies. The significantly different social activity variables include participation in gardening; golf; tennis; home entertaining; bridge clubs; book clubs; social, business, religious, and community organizations; and rock concert and play attendance.

A profile of bank card users

Based on an interpretation of the data, it is possible to develop a profile of bank credit card users based upon the weights of the standardized coefficients and upon group mean differences provided by the data in Tables 3 and 4.

First, with respect to store image characteristics as compared with nonusers, bank card users prefer to shop at stores with a greater variety of merchandise. They also show greater interest in credit plans available than do nonusers.

Second, the social activity variables lead one to conclude that users participate more in gardening, play more tennis, go to more

TABLE 4
CHARACTERISTICS DIFFERENTIATING BANK CARD
USERS FROM NONUSERS: TEST OF GROUP MEANS

Variables	Bank card nonusers	Bank card users	P
Credit or billing policies	1.790	1.920	.050
Merchandise variety	1.240	1.300	.100
Garden	.423	.504	.025
Golf	.081	.165	.001
Tennis	.213	.305	.010
Home entertaining	.478	.656	.001
Bridge clubs	.066	.133	.010
Rock concerts	.137	.086	.025
Plays	.163	.307	.001
Book Clubs	.121	.187	.010
Social organizations	.160	.260	.001
Community organizations	.118	.197	.010
Religious organizations	.307	.410	.010
Business organizations	.063	.138	.001

NOTE: F test.

plays, and engage more in home entertaining than do nonusers. Bank card users tend to belong to book and bridge clubs. They also appear to affiliate with religious, social, community, and business organizations to a greater extent than do nonusers. On the other hand, users are less likely to engage in photography as a hobby, are less likely to attend rock concerts or spectator sports, and are less likely to go camping.

Third, with respect to demographic variables, bank card users have a greater predisposition to possess credit cards and to establish credit at a variety of retail outlets than do nonusers. They tend to live in houses rather than in multiple unit dwellings. On the average, they have fewer children at home than do nonusers. Their spouses typically have more years of formal education. They have higher family incomes. Finally, bank card users are more likely than nonusers to be white and male.

Implications

This research has been undertaken with the aim of developing a valid description of bank credit card users which will be useful to regional chain department store retailers in devising marketing strategies to attract that segment of the market.

The profile can help to provide regional chain department store retailers and other regional and local retailers with insight into the social activities, store characteristic preferences, and demographic characteristics of bank credit card users. Information of particulat interest revealed by this research is the tendency of this market segment to use a variety of credit sources and to be interested in the credit plans available at different stores. This seems to imply that regional chain department stores and other local and regional retailers might attract bank card users by accepting bank credit cards as a vehicle for credit purchases and then switching these people to the store's own credit card or other credit accounts.

Several benefits might accrue to regional retailers who accept bank credit cards for charge purchases. First, newcomers who might otherwise continue to shop at their customary stores, that is, national department stores and specialty store chains, might be attracted more easily to the regional retailers. Second, the individual newcomer is essentially prescreened as a credit risk based on his or her current status as an active user of a bank card. This fact can reduce the cost of customer credit checks and perhaps reduce losses from unpaid accounts receivable. Third, by encouraging the bank credit card user to establish a credit account with the store, the retailer can expect to increase that customer's loyalty to the store.

In addition to these three benefits, a set of strategies also is suggested by the profile of bank card users presented in this study. The social activity characteristics of the users indicate that they have a pronounced interest in bridge clubs, tennis, gardening, and home entertaining. Regional retailers can develop strategies such as sponsoring organized clinics in gardening, bridge, and tennis to appeal to these interests. The regional retailer might also provide gourmet cooking classes and home decorating lessons. Such promotions will attract a high proportion of persons from the bank card user segment of the market. Furthermore, since bank card users also tend to be involved in formal organizations, such as church and civic groups, members of such organizations might be extended invitations to visit the store on a regular or special event basis.

Regional retailers also might attract this segment by selling theater tickets in their stores, as many do. They also might sponsor local theatrical productions and maintain a well stocked best-seller section in their book department.

SUMMARY

There appear to be three market segments which the regional or local retailer may be able to attract more effectively by accepting bank cards. These are newcomers, travelers, and persons wishing to carry a minimum of credit cards. All of these are likely to have bank cards. By catering to the life-styles, activities, and characteristics of bank card users, the local or regional retailer may increase penetration of these three segments.

This study has served two major purposes. First, it has facilitated the development of a descriptive profile of bank credit card users with respect to their store characteristic preferences and their social activities. The profile suggests several marketing strategies which could help regional retailers better compete with national chain department stores for the business of bank credit card users.

A second purpose has been to augment knowledge about a market segment of growing interest to marketing practitioners and academicians. This knowledge may further understanding of the value of segmentation as a marketing strategy.

REFERENCES

1. William O. Adcock, Elizabeth C. Hirschman, and Jac L. Goldstucker, "Bank Credit Card Users: An Updated Profile," paper presented at the Association for Consumer Research, 1976 Annual Conference, Atlanta, Georgia, 28–31 October 1976; R. V. Awh and

D. Waters, "A Discriminant Analysis of Economic, Demographic, and Attitudinal Characteristics of Bank Charge-Card Holders: A Case Study," *Journal of Finance* 29 (June 1974): 973–80; H. Lee Matthews and John W. Slocum, Jr., "Social Class and Commercial Bank Credit Card Usage," *Journal of Marketing* 33 (January 1969): 71–78; Joseph T. Plummer, "Life Style Patterns and Commercial Bank Credit Card Usage, *Journal of Marketing* 35 (April 1971):35–41; James B. Wiley and Lawrence M. Richard, "Application of Discriminant Analysis in Formulating Promotional Strategy for Bank Credit Cards," in *Advances in Consumer Research* vol. 2, edited by Mary Jane Schlinger (Urbana, Ill: Association for Consumer Research, 1974), pp. 535–44.

2. Awh and Waters, "Discriminant Analysis," found that most bank charge card users are also users of other types of credit cards, including those issued by department stores.

3. See "Bank Cards Push for the Big Stores," *Business Week*, 23 September 1976, p. 107.

4. Plummer, "Life Style Patterns and Card Usage."

5. Norman H. Nie, C. Hadlai Hull, Jean G. Jenkins, Karin Steinbrenner, and Dale N. Bent, *Statiscal Package for the Social Sciences*, 2d ed. (New York: McGraw-Hill Book Company:1975), p. 435.

6. The way to avoid the search bias is to develop discriminant functions on one-half of the data set and apply the obtained functions to the other half to test their validity. This method was used in the development of the profile for bank card users reported here. The total sample of 952 cases was divided into equal halves on an odd-even basis. A stepwise discriminant analysis using a combination of demographic, social activity, and store characterstics was carried out on the even-numbered half. The resulting sets of discriminant functions were then applied to the odd-numbered half for cross-validation. See R. E. Frank, W. F. Massey, and D. G. Morrison, "Bias in Multiple Discriminant Analysis," *Journal of Marketing Research* 2 (August 1965): 250–58.

4

Multivariate Analysis of Variance and Covariance

CHAPTER REVIEW

As theoretical constructs, Multivariate Analysis Of Variance (MANOVA) and Multivariate Analysis of Convariance (MANOCOVA) have been known for several decades since Wilks' original formulation [24] and Tukey's discussion of application [23]. However, it was not until the development of appropriate statistical tests with tabled values and the advent of high-speed computers and programs able to handle the complex computations that MANOVA became a practical tool for applied researchers. To assist in understanding our discussion of MANOVA and MANOCOVA, please review the Definitions of Key Terms.

Your understanding of the most important concepts in the use of MANOVA and Multivariate Analysis Of Covariance (MANOCOVA) should enable you to:
- [ ] Explain the underlying logic of ANOVA and its assumptions.
- [ ] Describe the purpose of "post hoc" (or multiple comparison) tests in analysis of variance.
- [ ] Discuss the role of experimental designs.
- [ ] Explain the differences among fixed effects, random effects, and mixed effects experimental designs.
- [ ] Explain what MANOVA does, and why it should be used instead of multiple ANOVA tests to compare two or more criterion variables per experimental treatment group.
- [ ] State the assumptions for use of MANOVA.
- [ ] Describe the purpose of multivariate analysis of covariance (MANOCOVA).
- [ ] Tell what Wilks' lambda statistic is used for in MANOVA.
- [ ] Identify the techniques that can be used to conduct post hoc or follow-up tests in MANOVA.

DEFINITIONS OF KEY TERMS

MATRIX. A rectangular, ordered array or table of numbers. These numbers are called the elements of the matrix. For example, an *m* by *n* matrix has *m* rows that each contain *n* elements.

DETERMINANT. A single number (derived in a precise way) to represent the value of a matrix. Symbolized by bars // placed around the matrix value (or determinant) which is computed as a function of the matrix's elements. Determinants are needed for calculating Wilks' lambda statistic.

WILKS' LAMBDA STATISTIC. The preferred statistic for testing the significance of each of the sources of variance in a MANOVA experimental design. It is used to test for the significance of the main (individual) effects and interaction (joint) effects of the independent variables on the dependent variables. It is simply a multivariate extension of the F-test in univariate ANOVA. Wilks' lambda is also referred to as the maximum likelihood criterion.

MULTIVARIATE NORMAL DISTRIBUTION. A generalization of a univariate normal distribution to the case of "p" variates, i.e., two or more dependent variables. A multivariate normal distribution of sample groups is a basic assumption required for validity of the significance tests used in MANOVA.

HYPOTHESIS EFFECTS. Differences among sample group centroids in multidimensional measurement space. The *null hypothesis* is that there are no differences among group centroids.

FACTOR. A nonmetric independent variable also referred to as a treatment or experimental variable.

TREATMENT. The independent variable that the researcher manipulates to see the effect (if any) on the dependent variables. The treatment variable can have several levels, e.g., different intensities of advertising appeals might be manipulated to see the effect on consumer believability.

MAIN EFFECTS. Individual effects of each independent variable on the dependent variable.

INTERACTION EFFECTS. The joint (or combination) effects of independent variables on the dependent variable beyond the separate influence of each independent variable.

EXTRANEOUS VARIABLE. Uncontrolled, non-manipulated independent variable which is merely a "nuisance" variable (that may confound the experiment) of no interest to the researcher. Covariate analysis can help eliminate this unwanted influence on the dependent variable.

COVARIATE ANALYSIS. Use of regression-like procedures to remove extraneous (nuisance) variation in the dependent variables due to one or more uncontrolled metric, independent variables. The covariates (uncontrolled independent variables) are assumed to be linearly related to the dependent variables. After correcting for the influence of the covariates, a standard ANOVA is carried out on the dependent variables. The procedure is much like the technique of multiple partial correlation adjustment.

SSCP MATRIX. The sums of squares and cross products matrix in MANOVA analogous to the sums of squares in ANOVA. It is obtained from the formula $W + B = T$ which is the multivariate version of the univariate $SS_w + SS_b = SS_T$, where the sums of squares within groups plus the sums of squares between groups equal the total sums of squares across all groups. Instead of single sums or numbers, MANOVA concerns itself with matrices or tables of numbers.

Λ **STATISTIC.**   Lambda statistic which is a test developed by Wilks for multivariate hypotheses. Compares the "within-groups" SSCP matrix to the "between-groups" SSCP matrix. The greater the true group differences in MANOVA, the smaller the value of lambda (Λ). Often referred to as Wilks' lambda.

CENTROID: A vector of means, or more simply stated, a mean value calculated using several other means as the raw data. In essence, a mean of means.

BIVARIATE: pertaining to two variables.

HYPOTHESIS: a statement or assumption about a population which the researcher attempts to accept or reject on the basis of statistical data about the population obtained from a sample.

ANALYSIS OF VARIANCE: a statistical technique used to determine if samples came from populations with equal means. Univariate analysis of variance employs one dependent measure, while multivariate analysis of variance employs two or more dependent measures to compare populations.

4

WHAT IS MULTIVARIATE ANALYSIS OF VARIANCE?

Like Univariate Analysis Of Variance (ANOVA), Multivariate Analysis Of Variance (MANOVA) is concerned with differences between groups (or experimental treatments). It possesses all the advantages of the univariate analysis in testing the effects of a number of independent variables in an experimental design. What really differentiates multivariate from univariate analysis of variance is that in MANOVA, the dependent variable is a vector (set or composite of measures) rather than a single mean response per group. Each treatment group is observed on two or more dependent (criterion) variables simultaneously, and the hypothesis tested is the equality of dependent mean vectors for the groups, as depicted in Figure 4.1

This dependent vector (or centroid) is presumed to have a multivariate normal distribution with the same within-group dispersion (or variance-covariance matrix) in each of the groups being studied. Assuming the equality of dispersion matrices is the MANOVA extension of the homogeneity of variance assumption in ANOVA designs. The research focus of MANOVA is whether or not there is a "real" difference (or differences) among the group centroids (mean vectors). Are some or all of the groups centered at different locations in the measurement space covered by the dependent vector variable? In Figure 4.2 a MANOVA design for a bivariate dependent (criterion) variable, where there are "real" differences among the groups, is illustrated.

RELATIONSHIP TO OTHER MULTIVARIATE TOOLS

Multivariate Analysis Of Variance (MANOVA) is a statistical technique which can be used to study the effect of multiple independent (treatment) variables measured on two or more dependent variables simultaneously. Similar to Multiple Discriminant Analysis (MDA) where the dependent variables are nonmetric, MANOVA is used when the independent variables are nonmetric. Covariance analysis can be applied in conjunction with MANOVA (the combination is called MANOCOVA) to remove after the experiment the effect of any uncontrolled independent variables on the independent variables. In MANOCOVA, there is a mixture of nonmetric and metric independent variables. In fact, all three (MDA, MANOVA, and MANOCOVA) may be considered special cases of canonical analysis (discussed in Chapter Five) in which either dependent or independent variables may be nonmetric.

127

UNDERLYING LOGIC OF ANOVA

Since MANOVA is merely an extension or generalization of ANOVA, and all the basic principles and techniques of experimental design and hypothesis apply to both, let us review our knowledge of univariate analysis of variance before we move to the multivariate case.

UNIVARIATE ANALYSIS OF VARIANCE

In its simplest form, ANOVA is concerned with the relationship between a single metric dependent variable and a single nonmetric independent variable at two or more treatment levels. Stated differently, ANOVA analyzes the effects of one or more categorical independent variables, measured at different levels, upon a continuous

Univariate Analysis

$H_0 = u_1 = u_2 = \ldots = u_k$

Null hypothesis (H_0) = all the group
means are equal, i.e., they come from
the same population.

Differences Among Group Means on one dependent (criterion) variable.

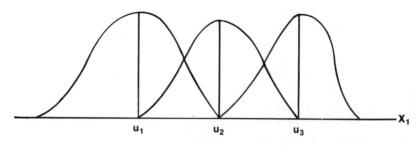

Figure 4.1—ANOVA Criterion Variables and Hypotheses

Multivariate Analysis

$$H_0 = \begin{bmatrix} u_{11} \\ u_{21} \\ \bullet \\ \bullet \\ \bullet \\ \bullet \\ u_{p2} \end{bmatrix} = \begin{bmatrix} u_{12} \\ u_{22} \\ \bullet \\ \bullet \\ \bullet \\ \bullet \\ u_{p2} \end{bmatrix} = \bullet\bullet\bullet = \begin{bmatrix} u_{1k} \\ u_{2k} \\ \bullet \\ \bullet \\ \bullet \\ \bullet \\ u_{pk} \end{bmatrix}$$

U_{11} = mean of variable 1, group 1

U_{12} = mean of variable 1, group 2

U_{21} = mean of variable 2, group 1

U_{22} = mean of variable 2, group 2

U_{pk} = mean of variable p, group k

Null hypothesis (H_0) = all the group centroids (mean vectors) are equal.

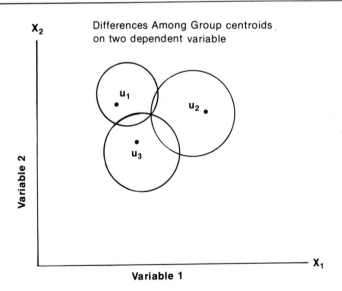

X_2

Differences Among Group centroids on two dependent variable

Variable 2

Variable 1

X_1

Figure 4.2—MANOVA Criterion Variables and Hypotheses

dependent variable assumed to be measured on an interval scale. For example, a researcher may want to determine what (if any) effect three different levels of advertising appeals (i.e., increasing levels of claims) for a product may have on the consumers' evaluations of the product's quality. Thus, the researcher would be studying the effect of one independent variable or experimental treatment (advertising appeals) at three different levels of intensity. The idea behind ANOVA is that the true population differences can be estimated from the sample groups . . . and that comparisons of these sample groups will reveal any real differences in the larger populations of interest. Our null hypothesis is always that the population means are equal. That is, if the population means are represented by the letter "U," then $U_1 = U_2 = U_3 = U_k$. From another viewpoint, if the null hypothesis is true, then each of three estimates of the population differences or variances should be equal (except for sampling error). These three estimates of population variance are each expressed as a "sum of squares", which is nothing more than a sum of squared deviations from a group mean.

1. Total variation or *total sum of squares* (SS_T) = sum of the squared deviations of each individual observation from the grand mean across all sample groups.
2. Between-group variation or *sum of squares between groups* (SS_B) = sum of the squared deviations of each group mean from the grand mean (adjusted for any differences in group size).
3. Within-group ("error") variance or *sum of squares within groups* (SS_W) = sum of the squared deviations of each observation from its own group mean.

As illustrated in Figure 4.3, the ANOVA process requires partitioning the total variance into two sources to form a ratio called the F-value:

$$F = \frac{\text{Sum of Squares Between-Groups [Variance Between Groups + Variance Within Groups (Error)]}}{\text{Sum of Squares Within-Groups [Variance Within Groups (Error)]}}$$

To compute the actual F-value, the SS_B and SS_W are divided by their appropriate degrees of freedom (the number of independent linear comparisons which can be made among the *n* observations) to derive a ratio of "mean squares" (MS). If there is no difference between treatment group means, then the numerator of the F-ratio (MS between groups) should be no larger than the denominator (MS within groups) except for sampling errors, i.e., an unrepresentative sample of the population. When the null hypothesis is not true, the between-

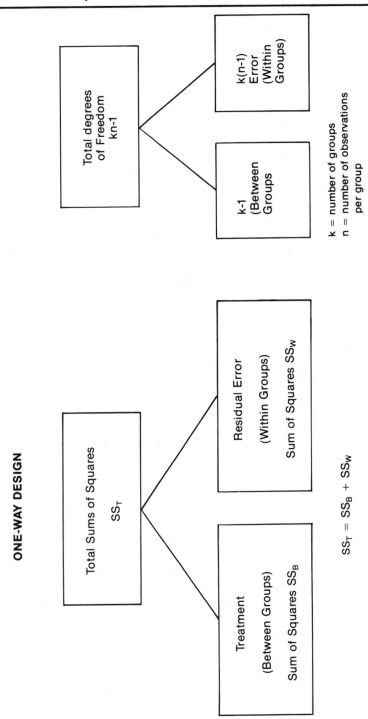

ONE-WAY DESIGN

Total degrees of Freedom
kn-1

k-1
(Between Groups)

k(n-1)
Error
(Within Groups)

k = number of groups
n = number of observations per group

Total Sums of Squares
SS_T

Treatment
(Between Groups)
Sum of Squares SS_B

Residual Error
(Within Groups)
Sum of Squares SS_W

$SS_T = SS_B + SS_W$

Figure 4.3—Partitioning of Variance Components

group mean square should be substantially larger than the within-group mean square as reflected in a high F-value. An F-test is carried out by comparing the computed F-value, for the appropriate degrees of freedom for the numerator and denominator, with tabled values of the F-distribution. If the F-table value is larger than the calculated F-ratio, any differences among the treatment groups are not significant. These tables show how often F-ratios of various sizes would occur if the null hypothesis is correct, given the assumption of samples taken at random from normally distributed populations with equal variances.

Because the F-ratio differs for each set of degrees of freedom and also for each level of significance, distribution tables are commonly available for only two levels of significance (.05 and .01). The reason that the .01 and .05 levels have come to be standard is largely due to the availability of tables for these levels, often borrowed from such disciplines as psychology where higher confidence levels are desired, not because these levels are sacrosanct. Decision makers in various functional areas often must take action with certainty of outcome levels much lower than .95 or .99.

ASSUMPTIONS OF ANOVA

It is desirable that certain conditions be met for proper application of analysis of variance. First, observations should be independent. That is, there should be a low correlation among them. This assumption is usually met by ensuring that samples are drawn randomly. Second, it is formally assumed that the variances are equal for all treatment groups, and that the populations from which they come are normally distributed. There is evidence that ANOVA is robust with regard to these assumptions, however [14]. Thus, the F-ratio may be relatively insensitive to violations of the assumptions where the sample sizes are about equal.

EXPERIMENTAL DESIGNS

ANOVA is particularly useful in carrying out experimental designs, i.e., research designs in which the researcher directly controls and manipulates one or more independent variables to determine the effect on a dependent variable. If nothing else causes the responses to vary, we can identify statistically significant differences by comparing variances due to the experimental treatment effect (between-groups variance) versus the measure of chance errors (within-groups variance). However, to conclude that the observed response (effect) was caused by the experimental manipulations, the experiment needs to be designed so as many extraneous factors as possible are eliminated. Examples of extraneous factors that must be controlled in a carefully planned experiment include [2, 5, 6]:

History—events external to the experiment but occurring simultaneously which might affect the response variable, e.g. a newspaper article on the dangers of saccharine appearing at the time a retailer attempts a price-cut on diet soft drinks.

Maturation—changes over time that internally alter the test subjects or units themselves, e.g., changes in consumption patterns of people as they grow older.

Testing—possible bias in response of test subjects due to awareness they are being observed, or due to prior exposure to test materials.

The primary function of an experimental design is to serve as a control mechanism (i.e. to control variance) so as to provide more confidence in concluding cause-and-effect relationships among variables. By developing an efficient design, the researcher seeks to (1) maximize the variance of variables of his research hypothesis, (2) control the variance of extraneous or "unwanted" variables that may affect results, and (3) minimize the error or random variance [2, 5].

To solve most problems, the researcher needs to understand only two levels of experimental design sophistication: (1) one-way and (2) factorial designs, i.e., two-way and higher designs [3, 20, 25].

One-Way Designs. The simplest design is one-way ANOVA which is used to investigate the effect of a single nonmetric independent variable (also called a factor) on a single dependent variable, where the independent variable has two or more levels (treatment or experimental groups). As a simple illustration, assume the HATCO regional sales manager is concerned that some of the older salespeople may by relying too much on repeat business from present customers instead of aggressively seeking new customers. It is decided to randomly select the names of eighteen salespeople, six each from one of three categories: (A) those with under 6 years service with the sales force, (B) those with 6-10 years service, and (C) those with over 10 years service. The sales manager writes a memorandum to HATCO's director of marketing research asking him to determine if there are significant differences among the three service categories with respect to the number of sales calls made on potential new customers during the last six months. Thus, there is one independent variable (years of service) with three levels of categories and one dependent variable (number of sales calls on potential new customers). Sales call information for

TABLE 4.1
SALES CALLS ON POTENTIAL NEW CUSTOMERS

Category A Salespeople	Category B Salespeople	Category C Salespeople
50	42	51
42	41	48
35	30	44
45	37	50
40	38	46
46	40	49
$\Sigma X_x = 258$	228	288
$M_s = 43$	38	48
$\Sigma X_t = 774$		
$M_t = 43$		

Note:
ΣX_x = sum of sales calls for each category of salesperson
M_s = average sales calls for each category of salesperson
ΣX_t = total sales calls for all categories of salespersons
M_t = average sales calls for all categories of salespersons

the three categories of sales force service are provided in Table 4.1, and the ANOVA summary showing an F-value of 8.5 in Table 4.2. Consulting a table of F values, we look at the category for 2 degrees of freedom (df) between samples (numerator) and 15 df within samples (denominator) to find a value of 3.68 at the .05 level. Our computed F-ratio of 8.5 is greater than the table value of 3.68 at the .05 level and even the 6.36 at the .01 level. Thus, there is a significant difference between the number of sales calls made by the three categories of salespeople during the past six months and the null hypothesis of no difference is rejected.

We know there is an overall significant difference among the three group means of 43, 38, and 48. But, is that significant difference between 48 and 38 only, or between 48 and 43 and 38 as well? To an-

TABLE 4.2
ONE-WAY ANOVA SUMMARY TABLE

Source of Variance	SS	df	MS	F-Ratio
Between Samples	300	2	150	8.5
Within Samples	264	15	17.6	

Total	564			

swer these questions, we need to follow-up with post hoc or multiple comparison tests (discussed in the next section).

POST HOC MULTIPLE COMPARISON TESTS

Although ANOVA can provide a method of rejecting the overall null hypothesis and accepting the alternative hypothesis that the independent sample means are not all equal, it does not pinpoint exactly where the significant differences lie. The researcher is not able to determine whether the null hypothesis (Ho) is rejected because $U_1 \neq U_2$; or because $U_2 \neq U_3$, or because all population means are unequal. Often, the analyst wants to determine exactly where the significant differences exist in the data.

When a statistical test is conducted, the researcher encounters two possible types of error risks: (1) rejecting a null hypothesis (that all the means are equal) when it is actually true (alpha "α" risk), and (2) failing to reject a null hypothesis when it is really false (beta "β" risk). It is important to control as much as possible the probabilities of making each of these two types of errors as depicted in Table 4.3.

Multiple t-tests are not appropriate for testing the significance of differences between the means of paired groups when a significant F-ratio has been obtained, since the probability of Type I error (i.e., rejecting the null hypothesis when it is actually true) increases with the number of intergroup means being tested for significance.[1] Several

---

[1] When only two independent sample means are being compared, ANOVA and the well-known t-test are equivalent since the F-ratio obtained with two sample means is simply the square of t, i.e., $F = t^2$, or $t = F^{1/2}$. Where several sample means need to be compared with one another, the researcher might be tempted to consider computing the t-ratio for the difference between all combinations of pairings. $N(N-1) \div 2$ is the formula which would indicate the number of t-tests required. For example, if five separate sample means (A,B,C,D,E) are being studied, ten t-tests will be required: $5(5-1) \div 2 = 10$ and the pairings would be as follows: A-B, A-C, A-D, A-E, B-C, B-D, B-E, C-D, C-E, D-E. The means are no longer completely independent since each group appears more than once in the paired t-tests. Not only are multiple t-tests the laborious way to identify specific significant differences among groups of data, but this procedure is questionable. As more t-tests are performed on correlated data, the probability that some t-tests will be statistically significant by chance alone increases. A researcher who decides to conduct multiple t-tests might select a traditional alpha significance level of decision rule such as .05. If he does, the probability of committing a Type I error (or obtaining a significant t-ratio) is greater than the desired .05 when the null hypothesis (Ho) that the samples come from the same population is true. In fact, the probability increases from .05 for two groups to .13 when all possible comparisons are made for three groups, .22 for four groups, .40 with five groups, etc. The ANOVA simultaneous test of significance avoids this bias since it determines whether differences within the entire distribution of sample means could have happened by chance, and thus whether the samples are actually drawn from the same general population.

options have been developed for making these comparisons: (1) Scheffé S test, (2) Tukey's honestly significant difference (HSD) method, (3) Tukey's extension of the Fisher least significant difference (LSD) approach, (4) Duncan's multiple range test, and (5) the Newman-Keuls test. These five post hoc or multiple comparison tests of significance between paired means have been contrasted for power by Winer [25]. He concluded that the Scheffé technique is the most conservative with respect to Type I error. The remaining tests were ranked in this order: Tukey HSD, Tukey LSD, Newman-Keuls, and Duncan.

Applying the Scheffé S test to find the confidence interval for the .05 and .01 levels of significance, we obtain the results reported in Table 4.4.[2] As can be seen, only Categories B and C are significantly different. Salespeople in Category C made significantly more sales calls on potential new customers than did salespeople in Category B. It would appear there may be a problem with Category B salespeople unrelated to years of service, since salespeople in both Categories A and C outperformed them.

---

[2] For a simple explanation of the procedures involved in carrying out the Scheffé S test, see H.H. Clarke and D.H. Clarke, *Advanced Statistics with Applications To Physical Education*. Englewood Cliffs: Prentice-Hall, 1972

TABLE 4.3
ERROR RISKS IN HYPOTHESIS TESTING

	Ho Actually True	Ho Actually False
	Correct Decision:	*Type II Error:*
Retain Ho	probability of retaining true Ho $1 - \alpha$	probability of retaining false Ho β
	Type I Error:	*Correct decision:*
Reject Ho	probability of rejecting true Ho α	probability of rejecting false Ho $1 - \beta$

Ho = Null Hypothesis

TABLE 4.4
SCHEFFÉ S TEST
ORDERED GROUP MEANS AND DIFFERENCES
BETWEEN GROUP MEANS AT THE .05 AND
.01 LEVELS OF SIGNIFICANCE

Means			Scheffé Test			
A	B	C	Difference	.05	.01	Significant
43	38		5	9.19	12.07	No
43		48	5	9.19	12.07	No
	38	48	10	9.19	12.07	p<.05

Factorial Design: Two-Way Analysis of Variance

Recall that one-way ANOVA designs involve a single nonmetric independent variable and a single metric dependent variable. In contrast to one-way designs, factorial or two-way ANOVA designs examine the effect (if any) of two or more nonmetric independent variables on a single metric dependent variable. As in one-way ANOVA, each of the independent variables may have two or more levels. Typically, there will be k levels (columns) of factor 1 and r levels (rows) of factor 2, which are displayed as shown in Table 4.5. With one-way ANOVA, the total variance is partitioned into the between-group variance estimate and the within-group variance (error) estimate. But in factorial designs, the between-group variance estimate itself is partitioned into (1) variation due to each of the independent variables (fac-

TABLE 4.5
DISPLAY OF TWO-WAY FACTORIAL DESIGN RESULTS

		Factor 1				Totals
		1	2	3 . . . . . k		
	1	X_{11}	X_{12}	X_{13}	X_{1k}	T_1
	2	X_{21}	X_{22}	X_{23}	X_{2k}	T_2
Factor 2	3	X_{31}	X_{32}	X_{33}	X_{3k}	T_3
		.	.	.	.	.
		.	.	.	.	.
		.	.	.	.	.
	r	X_{r1}	X_{r2}	X_{r3}	X_{rk}	T_r
Totals		T_1	T_2 . . . . T_k			T

TWO-WAY FACTORIAL DESIGN

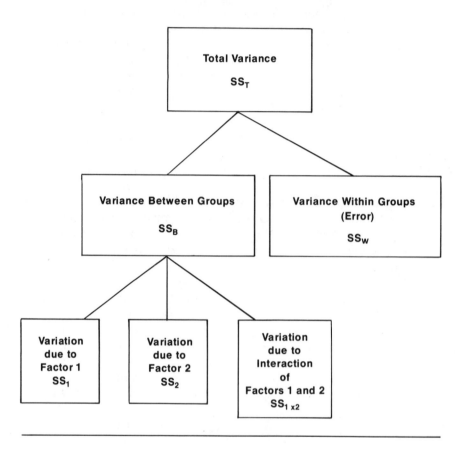

$$SS_T = SS_W + SS_B \text{ or } SS_T = SS_W + SS_1 + SS_2 + SS_{1 \times 2}$$

Figure 4.4—Partitioning Of Variance Components

tors) and (2) variation due to the interaction of the two variables—that is, their combined or joint effects on the dependent variable beyond the separate influence of each. Figure 4.4 illustrates this partitioning of variance for the two-way factorial design.

To obtain a better understanding of factorial designs, let us extend our earlier one-way example to the simplest type of factorial design— the two-way ANOVA. The HATCO regional sales manager has a hunch (hypothesis) that salespeople in better physical condition have more drive and confidence, and, therefore are most likely to make the highest number of sales calls on potential new customers. Let us assume each salesperson is classified on the basis of a medical examination into one of three categories of physical condition: (1) above average, (2) average, and (3) below average. The number of sales calls for the six salespeople in each category of physical condition (columns) and sales force service time (rows) are provided in Table 4.6. Note that this is a 3 × 3 factorial design since there are three levels for both independent variables, and that there are nine cells or possible combinations or interactions of salesperson service years and physical condition.

In two-way ANOVA problems, there are four sources of variance: (1) between columns, (2) between rows, (3) interaction, and (4) residual. *Residual variance* is the within-groups variance, and is the amount remaining after the other three sources of variance have been removed. Often, it is referred to as error variance since it consists of the unexplained, uncontrolled influences and serves as the denominator in all F-tests.

In order to determine the significance of the differences between the means of the k columns, r rows, and i interaction for the two factors of a two-way ANOVA, it is necessary to obtain three F-ratios. Thus, three null hypotheses are tested simultaneously by a two-way factorial design: (1) the effect of factor 1 on the dependent variable, (2) the effect of factor 2 on the dependent variable, and (3) the combined or joint effects of factors 1 and 2 on the dependent variable. Effects of the two independent variables (factors), individually, are called **main effects**; and their joint effect is called the **interaction effect**.

Results of our two-way ANOVA are provided in Table 4.7. Comparing our calculated F-ratios versus the tabled F-values shows all three sources of variance (service years, physical condition, and interaction of the two) significant beyond the .01 level. That is, this amount of variance would happen by chance less than one time in a hundred.

Again, our ANOVA has allowed us to reject the overall null hypothesis that all group means are equal. But we need to conduct "follow-up" tests to determine exactly where the significant differences lie between the groups. Applying the Scheffé test, we discover that there is a significant difference (.01 level) between salespeople in service years categories A and C, and also between B and C. Further, there are significant differences in the number of calls on potential new customers made by salespeople in physical condition categories A versus B, and A versus C. Those with over 10 years service make

TABLE 4.6
SALES CALLS ON POTENTIAL NEW CUSTOMERS BY SERVICE YEARS
AND PHYSICAL CONDITION

	Sales Force Service Years (Factor 1)			Row Sums for Physical Condition	Row Means for Physical Condition
	Under 6 yrs. Salesman A	6–10 yrs. Salesman B	Over 10 yrs. Salesman C		
Above Average	50	42	51		
	42	41	48		
	35	30	44		
	45	37	50		
	40	38	46		
	46	40	49		
	Σ 258	228	288	774	
	M 43	38	48		43
Average	40	39	36		
	38	36	39		
	34	38	34		
	37	36	39		
	34	41	36		
	39	38	32		
	Σ 222	228	216	666	
	M 37	38	36		37

Physical Condition (Factor 2)

Below Average	34	39	45	
	29	31	39	
	31	30	37	
	30	26	39	
	29	37	38	
	33	29	36	
	Σ 186	192	234	612
	M 31	32	39	34
Σ Column Sums for Service Years	666	648	738	2052 grand sum
M Column Mean for Service Years	37	36	41	38 grand mean

significantly more sales calls than do those in the two fewer years service categories. Also, those in above average physical condition make significantly more salescalls than those in average or below condition.

Interaction. We have determined that the variance attributed to the two factors operating separately (main effects) and jointly (interaction) is significant. Now it is necessary to remove the variance attributed to the factors separately to obtain the "pure" interaction between the factors. Variance due to "pure" interaction effects should only include variation which is not accounted for by either of the two main effects nor attributed to residual (error) variance. In other words, true interaction variance should be the non-error variance among the individual groups after the main effects variance has been removed.

Just as the main effects of sales force service years and physical condition can be independent of each other, so can the interaction effects be independent, which means that any combination of the three kinds of effects can prove significant. The interaction can be significant when none by itself, either one or the other by itself, or both of the factors combined are significant in their effect on the dependent variable (sales calls on potential new customers). Therefore, it is necessary to remove all differences among the means due to the main effects of the two factors to determine their true interaction effects. This too, can be accomplished by use of the Scheffé test and detailed explanations can be found in several texts [6, 16, 25].

Covariance Analysis. Designs in which metric independent variables are used with nonmetric independent variables are called "analysis of variance designs." The metric independent variables are referred to as "covariates," while the nonmetric independent variables are

TABLE 4.7
TWO-WAY ANOVA SUMMARY TABLE [1]

Source of Variance	SS	df	MS	F-Ratio	Tabled F(.01)
Columns: Service Years	252	2	126.0	10.33	5.11
Rows: Physical Condition	756	2	378.0	30.98	5.11
Interaction	288	4	72.0	5.90	3.77
Residual (error)	548	45	12.0		
Total	1,844				

[1] These results were calculated from the hypothetical data presented in Table 4.6.

termed "factors." Metric covariates are typically included in an experimental design to remove extraneous influences from the dependent variable, thus increasing measurement precision. Procedures similar to linear regression are employed to remove variation in the dependent variable due to one or more covariates. Then, a conventional analysis of variance is carried out on the "adjusted" dependent variable. Analysis of Covariance (ANOCOVA) is used in experimental designs where the researcher finds it impossible or awkward to control the assignment of treatments to subjects, or where background variables might affect responses from subjects.

Type of Effects. Another important consideration in experimental designs is the three different sampling situations to which the analysis of variance applies: (1) fixed effects, (2) random effects, and (3) mixed effects models. These models differ both in the way the experimental treatment groups are selected and in the kinds of inferences or interpretations that can be made from their analyses. The most common design is the *fixed effects* which assumes that treatment levels being studied are selected "a priori" (before-hand) and comprise the total population of interest. No attempt is made to generalize from these findings to other treatment levels. To illustrate, suppose HATCO wants to learn which method of compensation (salary or commission) provides more job satisfaction for the salesforce. Since the treatment groups (salary versus commission) are not randomly selected from a larger population of possible compensation methods, inferences must be restricted to the specific compensation categories in the experimental design. These two treatment categories are the only set of treatments about which we can make inferences . . . and each treatment is "fixed" in that it must appear in any complete replication of the experiment on new subjects (i.e., other samples of the salesforce).

Random effects designs assume the groups being studied are a random sample of possible treatment levels from a larger population with unknown mean and variance. Suppose it is hypothesized that the geographic size of a sales territory affects the salesperson's job satisfaction. Here, we can infer (or generalize) from our findings on the job satisfaction of salespeople from a random sample of sales territory sizes. Obviously, many other sales territory sizes are possible than actually occur as observations in the experiment, and HATCO is interested in the effect on job satisfaction of the whole range of possible levels (i.e., sales territory sizes). What we observe is just a random sample of territory sizes, but we generalize from these results to the entire population of territory sizes. That is, we can make inferences from our sample findings about the effect of territory size in general on job satisfaction. *Mixed effects* designs include one or more independent factors of each type: fixed and random [25].

SEPARATE VERSUS SIMULTANEOUS CONSIDERATION OF CRITERION VARIABLES

With respect to the independent (or treatment) variables, ANOVA has always been a "multi*variable*" technique since several "effects" (treatments or independent variables) are analyzed in terms of the change in the single dependent variable. But, there is always a single dependent (criterion) variable in univariate analysis.

A serious problem in ANOVA designs arises when the researcher attempts to consecutively measure two or more dependent (criterion) variables since no single criterion may represent all aspects of the problem, e.g., selection of the best advertisement from among several on the basis of such criteria as: attention, interest, believability, and product preference. Typically, where multiple criteria are studied, researchers select one or two dependent variables for decision making purposes leaving any remaining criteria for secondary consideration. Conflicting conclusions may be reached, even if the intercorrelations are ignored, when one of the advertisements is significantly superior on some of the variables and another advertisement is significantly superior on other variables. Thus, one has the dilemma: Which advertisement do you select to obtain the best overall combination of attributes? To overcome the limitations of univariate analysis, the researcher can turn to multivariate analysis of variance (MANOVA). Happily, there is a multivariate counterpart for every univariate analysis of variance research design.

MULTIVARIATE ANALYSIS OF VARIANCE

Instead of a single dependent variable, MANOVA examines the relationship between a combination of two or more dependent response measures, presumed to be metrically-scaled, and a set of predictor variables which are nonmetric (categorical). This simultaneous test for the effect on the combination of criterion variables is important because in most cases, the criterion variables are not really independent but correlated since they were obtained from the same individuals or subjects.

Prior to widespread use of high-speed computers for data analyses, multiple criterion variables were usually studied through repeat applications of univariate ANOVA until all the dependent variables had been covered. This approach, however, carries with it a Type I error tendency (outlined in Table 4.3) similar to that inherent in the use of multiple t-tests to pinpoint significance among three or more sample groups in ANOVA. Just as consecutive t-tests to determine significant differences on a single variable among multiple sample groups would be erroneous because of the increase in Type I error, so would a series of F-tests be inappropriate for sequentially testing the influence on multiple criterion variables among several groups.

MANOVA allows simultaneous testing of all the variables and considers the various interrelationships among them. It is particularly useful for examing interrelated criterion variables where individual ANOVA of each separate criterion might lead to erroneous conclusions. For example, although significant differences might be shown using sequential univariate analysis of each dependent variable, an overall (global) test of the multiple criteria simultaneously may not find significance. Conversely, as overall difference might be determined by MANOVA even though a series of univariate tests on each of the criterion variables uncovers no significant differences.

EXTENDING ANOVA TO MANOVA

Whereas ANOVA tests for differences among groups by employing the appropriate sums-of-squares, MANOVA uses the sums-of-squares cross-products (SSCP) matrices. Thus, the variance between groups is determined by partitioning the total SSCP matrix, then carrying out the appropriate tests of significance. Similar to the F-ratio of the between-group mean square to the within-group mean square in ANOVA, our multivariate F-ratio is merely a generalization to a ratio of the within-groups and total-groups dispersion (variance) matrices which tests for equality among treatment groups based on their respective centroids (vector means).

Basically, all the principles of ANOVA apply to MANOVA. Multivariate analysis of variance permits application of all the techniques of experimental design and hypothesis testing to multiple dependent variables. Beyond ANOVA however, MANOVA accounts for possible correlations among all dependent variables while testing all variables simultaneously. Further, it relieves the researcher from the job of selecting the most important dependent variable from several possible conflicting ones for decision making purposes.

MANOVA ASSUMPTIONS

Analogous to the assumption of common variance and normal distribution in ANOVA, it is assumed in MANOVA that the within-group dispersion matrices (representing averaged sums-of-squares and cross-products of deviations from their respective group means) are equal across the treatment groups, and that the set of dependent variables is multi-normally distributed.

MULTIVARIATE ANALYSIS OF COVARIANCE (MANOCOVA)

Use of multivariate analysis of covariance (MANOCOVA) with MANOVA improves the precision of an experiment by removing possible sources of variance in the criterion variable that may be accounted for by metric independent variables *not* controlled in the experimental design. Removing these extraneous influences reduces the residual error thereby increasing the "pure" effect of the treatment variables. Further, the analyst is not forced to match test and control groups on all background variables that might affect responses. Instead, MANOCOVA, which allows a mixture of nonmetric and metric-scaled predictors, employs the regression-like analysis to measure post hoc (after the fact), the effect of the uncontrolled predictor variables. This influence is removed or partialled out so that the effect of the controlled predictor variables in the experimental design can be measured independently. Thus, MANOCOVA derives a linear combination of the criterion variables whose variance is best explained by the treatment variable after removing the effect of any covariates.

MANOCOVA can be viewed as MANOVA on multiple regression residuals (multiple regression residuals can be defined as the residual variance not explained by the metric independent variables). Its real value is evident when there is high correlation between the covariates (metric treatment variables) and the criterion variables. It permits post experiment statistical control on one or more extraneous variables by removing their influence (from group comparisons) on the main treatment variable. As a word of caution, however, this tool is effective only when the relationship between the covariates and the criterion variables is linear, and where the extent of this relationship is not dependent upon the nonmetric independent (treatment) variables.

Review of the periodical literature readily reveals that applications of MANOVA and MANOCOVA to behavioral science problems have been minimal to date [8, 9, 19, 26]. But their use will surely increase as researchers learn more about the applications and value of these techniques for measuring multiple criterion variables.

HATCO SALES TRAINING PROBLEM

Having reviewed the underlying logic, assumptions, and limitations of ANOVA, MANOVA, and MANOCOVA in this chapter, let us make use of our Chapter One data bank to work out some sample problems. First, we will conduct a one-way MANOVA experiment. Subsequently, we will carry out a two-way (factorial) MANOVA experimental design. Since this book is oriented toward data analysts instead of statisticians, our discussion in solving the sample problems will focus on conceptual understanding and the explanation of the output of computer programs.

MANOVA and MANOCOVA are typically used in experimental designs where an equal number of responses (observations or replications) can be obtained from each combination of treatment levels. When an equal number of responses is obtained, the design is referred to as having equal "cell" sizes. When the number of responses from each combination of treatment levels is *not* equal, it is referred to as having unequal cell sizes. Designs with equal cell sizes are called orthogonal designs. Without orthogonality, the computational formulas for MANOVA and MANOCOVA increase sharply in complexity. However, procedures do exist to handle unequal cell sizes [19]. For illustration simplicity, we will assume that the non-metric independent variables (factors) in our experimental design are "fixed." Recall that when the independent factors constitute the entire population of research interest, the design has *fixed-effects* and no effort is made to infer beyond the groups being analyzed. In both our one-way and two-way MANOVA examples, we will first show results employing no covariates, then all six covariates (or extraneous independent variables)—social psychological attitudes tests—from our data bank in Chapter One.

One-Way MANOVA Problem

We shall use our HATCO data bank to set up the problem. Recall that one-way MANOVA designs involve a single nonmetric independent variable and two (or more) metric dependent variables. Let's assume the following. HATCO wants to test the effectiveness of two different methods of sales training for its salespersons. The two methods, which will supplement regular reading assignments and classroom lectures, are (1) role playing and (2) case study analysis. The two methods of training represent the nonmetric independent variable (treatment) for our experiment. Note that since the treatment levels or types being studied (role playing versus case study) are selected prior to the experiment, this represents a fixed effects design.

If we measured the effectiveness of each method by a single score on each salesperson, the design would be a one-way univariate analysis of variance (ANOVA). To convert our problem into a one-way MANOVA design, we must have two separate measurements on the effectiveness of each method (i.e., two dependent variables). For our example, we will assume the two dependent measures are variables X_7 and X_8 from our data bank. Variable X_7 measures overall knowledge about the company's products, and Variable X_8 measures motivation toward achievement in the individual's sales career.

In setting up our design, two additional considerations need to be pointed out. To simplify our problem we would like to have equal cell sizes. Our data bank has observations on 50 salespersons and equal cell sizes are possible for a one-way MANOVA ($50 \div 2 = 25$). But, for a two-way MANOVA 50 observations will not permit equal cell sizes ($50 \div 4 = 12.5$). Therefore, to obtain equal cell sizes we will assume that two vacationing salespeople were unable to participate in the experiment—the result is a sample of 48 salespersons. Using the reduced sample of 48 observations, each salesperson is randomly assigned to one of two treatment groups, i.e., the role playing or the case study group. This procedure ensures that the method of selecting individuals for either the case study or role playing groups does not influence the results of our experiment.

The design we have set up for HATCO enables us to determine which (if either) training method will yield a significantly higher composite score on the two dependent measures of effectiveness—variables X_7 and X_8. The null hypothesis (where "u" is a centroid or vector of means for the dependent variables) is H_0: $u_1 = u_2$; the alternative hypothesis is H_1: $u_1 \neq u_2$. Scores obtained by the salespeople on the two tests are graded on a scale from 1 to 15 (low to high) carried out to three decimal places. The test scores, by sales training method, for each of the 48 salespeople are displayed in Table 4.8.

Significance Testing. From the total sums of squares and cross products matrix (SSCP), a test is conducted to determine the significance of differences (if any) among treatment groups. Wilks' lambda (Λ), sometimes referred to as the U statistic, is the general statistic used to test for centroid equality in the case of several groups with multiple criterion variables. Special cases of Wilks' lambda (Λ) for two groups are [22]:

1. *Hotelling's* T^2—an extension of the univariate t-test to the multivariate case.
2. *Mahalanobis' D^2*—a measure of generalized distance between two groups which considers the covariation among the variables generating the distance calculation.

Hotelling's T^2 and Mahalanobis' D^2 statistics are not applicable to the analysis of multivariate responses when there are three or more treatment groups. In such cases, Wilks' lambda or Bartlett's V and Rao's Ra functions must be computed to test for group differences [22]. There are only two groups in our experiment. But Wilks' lambda is still appropriate since it can be used with only two groups, or with three or more. It also is a good test statistic to use since it is the one most likely to be presented in computer programs. Its formulation is: $\Lambda = \frac{|W|}{|W + A|}$ where /W/ is the determinant (simply a single number, derived in a precise way, to represent the value of a square matrix of numbers) for the SSCP matrix for within groups. /W + A/ is the determinant for the total SSCP matrix formed by adding the corresponding elements of the *within groups* SSCP matrix and the *between (among) groups* SSCP matrix for the hypothesis under investigation (i.e., the equality of treatment groups).

Each F-value in ANOVA has an analogous Wilks' lambda statistic in MANOVA. As we have learned, Wilks' lambda is simply the ratio of two determinants /W/ ÷ /T/ where /W/ is the determinant of the within-groups sums-of-squares and cross-products, and /T/ is the determinant of the total groups sums-of-squares and cross-products. The smaller the value of the lambda statistic, the greater the implied statistical significance among the group centroids. This is because, in our ratio, /W/ becomes smaller relative to /T/ or /W + A/ when the variance among the group centroids is relatively larger than the within groups variance. In cases involving several (dependent variables) and several treatment groups, the distribution of the lambda statistic is quite complex and requires use of some approximation to it for testing significance [22].

For any number of dependent variables (p) and up to three independent variables (k) or for any number of predictor variables and two criterion variables, the distribution of the lambda statistic precisely matches the F-distribution. In our problem, where there are two dependent variables (p = 2) and two treatment groups (k = 2), an exact significance test is available, so no approximate test statistic is required.

TABLE 4.8
SALESPERSONS' TEST SCORES BY TRAINING METHOD
(Basic Data[1] for One-Way MANOVA: $p = 2$, $k = 2$, $n = 24$)

Treatment 1 = Role Playing			Treatment 2 - Case Studies		
Test Scores	Knowledge X_7	Motivation X_8	Test Scores	Knowledge X_7	Motivation X_8
Salesperson			Salesperson		
1.	7.952	7.473	25.	8.599	10.364
2.	10.723	10.724	26.	10.222	9.836
3.	8.911	8.817	27.	9.134	11.190
4.	7.113	7.321	28.	9.766	9.868
5.	9.652	10.369	29.	11.608	13.504
6.	8.583	6.654	30.	10.547	11.792
7.	7.688	6.434	31.	7.532	9.988
8.	8.812	8.527	32.	12.910	11.911
9.	5.865	7.816	33.	9.421	9.945
10.	8.149	8.245	34.	11.869	12.032
11.	8.030	6.820	35.	10.787	11.680
12.	6.937	8.112	36.	7.771	10.959
13.	6.362	7.752	37.	6.144	7.999
14.	7.525	7.433	38.	8.294	9.996
15.	10.041	9.914	39.	11.337	12.120
16.	10.257	11.014	40.	8.118	8.948

	X₇	X₈		X₇	X₈	Marginals
17.	5.581	6.613	41.	10.879	9.604	433.400
18.	9.148	10.157	42.	11.924	12.120	451.501
19.	8.826	9.653	43.	10.618	7.500	9.029
20.	8.210	8.169	44.	7.724	7.316	9.406
21.	7.625	8.838	45.	12.066	10.198	4,063.319
22.	7.457	8.854	46.	10.865	11.841	
23.	9.489	8.474	47.	9.402	10.503	4,383.730
24.	10.568	10.017	48.	6.359	8.501	4,153.872
Totals	199.504	204.200		233.896	247.301	
Means[2]	8.313	8.508		9.746	10.304	
$(\Sigma X_7)^2$	1,703.946			2,359.373		
$(\Sigma X_8)^2$	1,779.871			2,603.859		
$\Sigma X_7 X_8$	1,703.466			2,450.406		

[1] The basic data symbols are: p = number of dependent variables, k = number of independent variables, and n = treatment group size.

[2] The *Group Centroids* are displayed as: $T_1 = \begin{bmatrix} 8.313 \\ 8.508 \end{bmatrix}$ $T_2 = \begin{bmatrix} 9.746 \\ 10.304 \end{bmatrix}$

Are the centroids of these two treatment groups significantly different?

MANOVA COMPUTER PROGRAM

Data analysts can use one of several available computer programs to carry out MANOVA [3, 10, 11]. Two of the more popular ones are the BMD [10] and the SAS [3] packages. The summary output from the BMD-12V MANOVA program, which can be used for both univariate and multivariate analysis of variance and covariance, looks like the format in Table 4.9. Our two dependent variables (X_7, product knowledge and X_8, motivation) are identified at the top, followed by the source of centroid differences, i.e., factor A (sales training methods), the log of the generalized variance which is the natural logarithm of the determinant for each SSCP matrix (analogous to a mean square in ANOVA), the U-statistic or Wilks' lambda statistic, the degrees of freedom associated with lambda, and the approximate F-statistic with its degrees of freedom.

With 2 and 45 degrees of freedom, the tabled F-value at the .05 level of confidence is 3.20 and the .01 level is 5.11. Our MANOVA test indicates that the mean vectors of the two treatment groups are not equal since our computed F-ratio (8.9) is much larger than the values from the F-table. Therefore, we reject the null hypothesis. Inspection of our treatment means from Table 4.8 shows that treatment two, case study analysis, is a significantly better training method than treatment one, role playing.

TABLE 4.9
MANOVA SUMMARY TABLE

BMD 12 V
MANOVA
Dependent Variables   7   8

Source	Log (Generalized Variance)	U-Statistic	Degrees of Freedom	Approx. F Statistic	Degrees of Freedom
A	9.17708	0.716353	2,1,46	8.9091	2,45
S(A)	8.84350				

Note: *Source of Variation*
  A = sales training methods (role playing vs. case studies)
  S(A) = residual error or variance due to individual unexplained differences among salespersons
  Dependent Variables
  7 = product knowledge
  8 = motivation of salesperson toward achievement in sales

POST HOC TESTS

In cases where there are three or more dependent variables for which overall (global) significance has been determined, we may want to learn which of the independent variables contributed to the overall significance. Several approaches may be considered [21, 22]. First, a series of univariate F-tests can be conducted on the individual dependent variables [8]. This approach, however, has been sharply criticized since it overlooks possible correlations among the dependent variables. Second, simultaneous confidence intervals can be computed for each of the variables [17], but this method is appropriate only for two groups. Further, it is an especially conservative technique which ought to be employed only when the cost of a Type I error is high. Third, a procedure similar to stepwise regression called "step-down analysis" [21] may be applied. It provides for computing each successive F-value only after eliminating the effects of the previous dependent variables. If the variables are uncorrelated, this step-down approach is really just a series of univariate F-tests. Finally, multiple discriminant analysis (MDA) may be applied to the SSCP matrices associated with significant effects to determine which variables contributed significantly to the overall MANOVA significance. The relative importance of each independent variable can be identified by deriving correlations between each original dependent variable and the discriminant function [21, 22].

> One of the by-products of a one-way MANOVA is the set of coefficients which produce the largest possible univariate F-ratio when they are taken as the coefficients of a linear function of the original variables. This linear combination of the original p variables is the discriminant function (more accurately, the first or primary discriminant function) for these k groups, and is a logical candidate for use in classifying some new subject whose classification is unknown into one of the k groups [14, p. 108].

A test of the significance of a multiple discriminant function is much like a test of centroid equality, except that the role of independent and dependent variables are reversed. MANOVA and MDA are actually closely related techniques which can complement each other in data analysis. Discriminant analysis seeks to determine the coefficients of linear combinations of the variables which best discriminate between multiple groups. In other words, the purpose is to identify the coefficients that maximize between-group variance with respect to within-group variance for hypothesized groups. MANOVA, on the other hand, tests whether already identified groups are significantly different. When a difference is discovered, discriminant analysis can be used to pick out those variables which best differentiate between the groups.

ONE-WAY MANOCOVA

Although it appears that case study analysis is superior to role playing in training salespeople, HATCO management is concerned about the possible influence that prior knowledge may have had on the findings. All new hired salespeople are given a series of six social-psychological attitude tests during the first year of their employment. Fearing the possible influence of these prior tests on the response variables in our just completed experiment, management asks if the influence of these extraneous variables can be removed after the fact, i.e., after the completion of our experiment. "Yes," the researcher says, "there's a technique called multivariate analysis of covariance (MANOCOVA) that can remove this influence ex post facto from our criterion variables . . . and I will apply it."

Multivariate analysis of covariance (MANOCOVA) is accomplished by designating one of the sets of uncontrolled, extraneous variables (i.e., a pre-experiment social-psychological test score) as a set of independent design variables. To illustrate, in our problem we now have (1) a set of metric criterion variables, (2) a set of nonmetric predictor variables, and (3) a set of metric, moderator or control variables (i.e., covariates). Our computer program can again be used to perform MANOCOVA this time to determine whether the treatment groups show centroid differences on the criterion variables once the effects of the covariates are removed.

In our data bank, covariates X_1 through X_6 are the scores on the various social-psychological tests given to each HATCO employee shortly after being hired. To remove the influence of these prior tests on our criterion variables, we rerun the BMD-12V program designating variables X_1 through X_6 as covariates. Results would appear like those shown in Table 4.10. Note that covariates X_3 and X_4 have a very significant influence on the two criterion variables . . . and, along with the other covariates, have reduced the main effect (training methods) F-ratio to 3.02, just below the .05 level of significance. Once the influence of all six covariates are removed or partialled out, the treatment centroids are no longer significantly different. Thus, the training methods are not significantly different when adjusted for background social-psychological variables.

Covariates, or extraneous background variables, are usually of little interest to the researcher for interpretative purposes. They are viewed as nuisance variables whose effects on the criterion variables need to be removed to increase the precision of the experiment in determining the main experimental treatment effects.

TWO-WAY MANOVA EXAMPLE

Continuing with the same data from our one-way MANOVA, we can convert it to a two-way MANOVA problem by adding another non-metric independent, experimental design variable. This is accomplished by random assignment of salespeople to two additional groups. Assume HATCO management wants to know not only whether one of the two

TABLE 4.10
MULTIVARIATE ANALYSIS OF COVARIANCE

Dependent Variables   7   8
Covariates   1   2   3   4   5   6

Source	Log (Generalized Variance)	U-Statistic	Degrees of Freedom	Approximate F-Statistic	Degrees of Freedom
A	6.83030	0.865756	2,1,40	3.0237	2,39
Covariate 1	6.71069	0.996344	2,1,40	0.0716	2,39
Covariate 2	6.69533	0.998629	2,1,40	0.0268	2,39
Covariate 3	15.19878	0.439914	2,1,40	24.8269	2,39
Covariate 4	10.97637	0.609141	2,1,40	12.5123	2,39
Covariate 5	6.99651	0.955641	2,1,40	0.9051	2,39
Covariate 6	7.63439	0.875794	2,1,40	2.7655	2,39
S(A)	6.68615				

Note: *Sources of Variation*
   A = sales training methods (role playing vs. case studies)
   Covariates = scores on various social-psychological tests given to each HATCO employee upon employment
     1 = self-esteem
     2 = locus of control
     3 = alienation
     4 = social responsibility
     5 = Machiavellianism
     6 = political opinion
   S(A) = residual error or variance due to individual unexplained differences among salespersons
     Dependent Variables
         7 = product knowledge
         8 = motivation of salesperson

TABLE 4.11
SALESPERSONS' TEST SCORES BY TRAINING METHOD AND BY TIME OF DAY
(Basic Data for Two-Way MANOVA, $p = 2$, $k = 2$, $n = 12$)

Factor A—Mornings

	Treatment 1 - Role Playing			Treatment 2 - Case Studies	
Test Scores	Knowledge X_7	Motivation X_8	Test Scores	Knowledge X_7	Motivation X_8
Salesperson			Salesperson		
1.	7.952	7.473	25.	8.599	10.364
2.	10.723	10.724	26.	10.222	9.836
3.	8.911	8.817	27.	9.134	11.190
4.	7.113	7.321	28.	9.766	9.868
5.	9.652	10.369	29.	11.608	13.504
6.	8.583	6.654	30.	10.547	11.792
7.	7.688	6.434	31.	7.532	9.988
8.	8.812	8.527	32.	12.910	11.911
9.	5.865	7.816	33.	9.421	9.945
10.	8.149	8.245	34.	11.869	12.032
11.	8.030	6.820	35.	10.787	11.680
12.	6.937	8.112	36.	7.771	10.959
Totals	98.415	97.312		120.166	133.069
Means	8.201	8.109		10.014	11.089
$(\Sigma X_7)^2$	825.211			1232.997	
$(\Sigma X_8)^2$		809.471			1489.988
$\Sigma X_7 X_8$	810.926			1346.023	

$G_7 = 218.58$
$G_8 = 230.38$

$M_7 = 9.11$
$M_8 = 9.60$

Factor B—Afternoons

13.	6.362	7.752	37.	6.144	7.999
14.	7.525	7.433	38.	8.294	9.996
15.	10.041	9.914	39.	11.337	9.706
16.	10.257	11.014	40.	8.118	8.948
17.	5.581	6.613	41.	10.879	9.604
18.	9.148	10.157	42.	11.924	12.120
19.	8.826	9.653	43.	10.618	7.500
20.	8.210	8.169	44.	7.724	7.316
21.	7.625	8.838	45.	12.066	10.198
22.	7.457	8.854	46.	10.865	11.841
23.	9.489	8.474	47.	9.402	10.503
24.	10.568	10.017	48.	6.359	8.501
Totals	101.089	106.888		113.730	114.232
Means	8.424	8.907		9.478	9.519
$(\Sigma X_7)^2$	878.735			1126.376	
$(\Sigma X_8)^2$		970.400			1113.871
$\Sigma X_7 X_8$	919.540			1104.383	

$G_7 = 214.82$
$G_8 = 221.12$

$M_7 = 8.95$
$M_8 = 9.21$

training methods is more effective (as measured by the combination of criterion variables), but also whether providing the training in the morning hours as opposed to the afternoon hours would affect test performances of the salespeople. Now, we must study two independent variables: (1) training method, and (2) time of training session. Our experimental design becomes a 2 × 2 matrix of independent variables, i.e., the intersection of columns (method of training) and rows (time of training)—each with two response measures. This design is called a *full factorial design* because the responses are studied for all combinations of two or more factors, each at two or more levels.

Our basic data for this two-way MANOVA is presented in Table 4.11. We assume each of the salespeople has been assigned to one of the cells in a systematic random manner. Cell means and marginal means are shown for twelve replications (observations) each for the four combinations of factors A and B with treatments 1 and 2.

Recall that in a two-way experimental design, there are four sources of variance: (1) between columns or treatments, (2) between rows or factors, (3) interactions between columns and rows, and (4) residual or error. Residual variance is the within-groups variance which is the amount remaining after the other three sources of variance have been removed. Often, it is referred to as error variance since it consists of the unexplained, uncontrolled influences and serves as the denominator in all F-tests.

To determine the significance of the differences among the centroids of the columns, rows, and interaction between the two independent factors of a two-way MANOVA, it is necessary to obtain three F-ratios. Thus, three null hypotheses are tested simultaneously by a two-way factorial design: (1) the effect of factor A on the criterion variables, (2) the effect of factor B on the criterion variables, and (3) the combined or joint effects of factors A and B on the criterion variables. The effects of the two independent factors (training methods and time of day), individually, are called *main effects* and their joint effects are called the *interaction effects*.

Assumptions of MANOVA

Remember several conditions must be met for the proper application of the multivariate analysis of variance. First, the residual or error variance should be normally distributed. That is, the population from which the error variance is figured should come from multivariate normally distributed universes since it represents the denominator in the calculation of the F-ratio. Second, error variance should be equal among the cells. That is, the variance-covariance matrices within each group should be approximately equivalent. Third, observations within cells should be independent—there should be a low correlation among them. This assumption is usually met by insuring that sample groups are drawn randomly.

MANOVA AND MANOCOVA COMPUTER PROGRAM

Turning again to the results from our BMD-12V computer program, the MANOVA analysis shown in Table 4.12 reveals a highly significant F-value of 10.5 for source of variance A (sales training methods). There is no significant difference in test scores by factor B (time of day of the training), but there is a significant interaction (AB) effect, i.e., there is a significant difference by sales training method by time of day. Reviewing the cell means for criterion variables X_7 and X_8 in Table 4.11, we can see that treatment two (case study analysis) yields higher scores than treatment one (role playing), and that factor A (mornings) gives higher scores than factor B (afternoons). Further, the highest scores are recorded when case study analysis is used in classes conducted in the morning.

In our simple two group case, it is not difficult to pick the best performing variable when significance results are found. But, in cases where you are studying more than two groups, it is often desired to

TABLE 4.12
MULTIVARIATE ANALYSIS OF VARIANCE

BMD12V
Dependent Variables   7   8

Source	Log (Generalized Variance)	U-Statistic	Degrees of Freedom	Approxi-mate F Statistic	Degrees of Freedom
A	8.97196	0.671915	2,1,44	10.4981	2,43
B	8.60162	0.973091	2,1,44	0.5945	2,43
AB	8.82207	0.780570	2,1,44	6.0440	2,43
S(AB)	8.57434				

Note: *Sources of Variation*
  A = Sales training methods (role playing vs. case studies)
  B = time of day (mornings vs. afternoons)
  AB = interaction between sales training methods and time of day
  S(AB) = residual error or variance due to individual unexplained differences among salespersons
    Dependent Variables
      7 = product knowledge
      8 = motivation of salesperson

pinpoint differences between groups when overall significant results
are obtained in MANOVA. The preferred approached to more detailed
analysis via multiple comparison tests include [12, 18, 21, 22]:

1.   Conducting multiple discriminant analysis (MDA) on the SSCP
     matrices associated with significant effects. MANOVA is used
     to test whether groups already identified are significantly dif-
     ferent. When a difference is found, discriminant analysis may
     be employed to determine the major areas of differences
     between the groups. One can also compute structure correla-
     tions or product moment correlations between each original

TABLE 4.13
MULTIVARIATE ANALYSIS OF COVARIANCE

| Dependent Variables   7   8 | | | | | |
| Covariates   1   2   3   4   5   6 | | | | | |
Source	Log (Generalized Variance)	U-Statistic	Degrees of Freedom	Approxi- mate F Statistic	Degrees of Freedom
A	6.59125	0.847561	2,1,38	3.3273	2,37
B	6.46004	0.966391	2,1,38	0.6434	2,37
AB	6.65578	0.794598	2,1,38	4.7822	2,37
Covariate 1	6.47539	0.992350	2,1,38	0.1426	2,37
Covariate 2	6.44441	0.997121	2,1,38	0.0534	2,37
Covariate 3	15.58779	0.412237	2,1,38	26.3771	2,37
Covariate 4	10.54963	0.609107	2,1,38	11.8723	2,37
Covariate 5	6.73456	0.954161	2,1,38	0.8888	2,37
Covariate 6	7.78229	0.825703	2,1,38	3.9052	2,37
S(AB)	6.42586				

Note: *Sources of Variance*
   A = Sales training methods (role playing vs. case studies)
   B = time of day of the sales training (morning vs. afternoon)
   AB = interaction of sales training methods with time of day
   Covariates = scores on various social-psychological tests given to each HATCO
      employee upon employment
   1 = self-esteem
   2 = locus of control
   3 = alienation
   4 = social responsibility
   5 = Machiavellianism
   6 = political opinion
   S(AB) = residual error or variance due to individual unexplained differences
      among salespersons
   Dependent Variables
   7 = product knowledge
   8 = motivation of salesperson

dependent variable and the discriminant function to identify the relative importance of the predictor variables.

2. Considering each of the criterion variables separately in a univariate ANOVA to determine the F-ratios. Such univariate ANOVA's are not independent so a procedure of stepdown F-ratios can be used by arranging the criterion variables in some a priori order of assumed importance to the problem, then running a sequence of F-tests. Through covariance, each F-ratio is computed with the effects of the preceding variables progressively removed.

Our MANOVA results identified one significant main effect and a significant interaction effect. But will these effects remain significant if MANOCOVA is applied using the six social-psychological measures as covariates? Stated another way, if the influence of the six social-psychological measures (covariates) on the dependent variables is removed, will the treatment centroid still be significantly different? The results of MANOCOVA with the six covariates are shown in Table 4.13.

Even when the effects of all six covariates are removed, there are still significant differences between levels of factor A and in the AB interaction. However, covariates 3, 4, and 6 were highly significant in accounting for the criterion variables. As mentioned earlier, the covariates themselves are seldom of interest to the researcher for interpretation since they are generally viewed as nuisance factors interfering with the precision of the experiment until their influence is partialed out. Generally, it is assumed that the covariate-by-factor (independent variable) interaction is zero, and this assumption is typically tested prior to analysis of covariance.

SUMMARY

We have not covered all the various types of experimental designs for MANOVA and MANOCOVA, nor would such an undertaking be practical in a book oriented toward the non-statistician. However, it should be noted that the MANOVA model is capable of extension in all the directions for which univariate analysis of variance can be applied. MANOVA is a valuable generalization of ANOVA for data analysts because it permits the simultaneous testing of all variables, accounts for any correlation among the variables and makes it unnecessary to make a priori decisions regarding which variables are most important. It is hoped the reader has obtained sufficient stimulation, understanding, and confidence to tackle some of the more sophisticated designs of MANOVA and MANOCOVA described in more statistically oriented texts [1, 6, 7, 12, 17, 20, 22, 25].

END OF CHAPTER QUESTIONS

1. Design a two-way MANOVA experiment. Name some extraneous factors that should be controlled in your experiment, e.g., history, maturation, testing effects, etc. What are the different sources of variance in your experiment? Are your treatment effects fixed, random, or mixed?

2. Locate and compare at least two "canned" computer programs for multivariate analysis of variance and covariance. What are the essential differences between the programs, especially with regard to the printout of results?

3. After finding overall or global significance, there are at least three approaches to doing follow-up comparison tests: (a) a series of univariate tests on each of the individual response variables, (b) stepdown analysis similar to stepwise regression in that each successive F-statistic is computed after eliminating the effects of the previous response variables, and (c) multiple discriminant analysis on the SSCP matrices associated with significant effects to determine the major differences among the groups. Try to name the advantages and disadvantages of these three approaches to pinpointing significant differences after finding overall significance among the groups.

4. Describe some data analysis situations in which multivariate analysis of variance and covariance would be appropriate in your areas of interest. What type of uncontrolled variables or covariates might be operating in each of these situations?

5. How do you believe the increasing use of multivariate analysis techniques will change the discipline with which you are most identified? Do you think studying phenomena under conditions approaching their natural complexity is a more realistic goal in some disciplines than others? Explain your answer.

REFERENCES

1. Anderson, T. W., *Introduction to Multivariate Statistical Analysis.* New York: Wiley and Sons, 1958.
2. Banks, S. *Experimentation in Marketing.* New York: McGraw-Hill, 1965.
3. Barr, A. J. et al (eds.) *A User's Guide to SAS 76.* Raleigh, North Carolina: Sparks Press, 1976.
4. Bock, R. D., "Contributions of Multivariate Experimental Designs to Educational Research," in R. B. Cattell (ed.), *Handbook of Multivariate Experimental Psychology* (Chicago: Rand McNally, 1966), pp. 820–840.
5. Campbell, D. T. and J. C. Stanley, *Experimental and Quasi-Experimental Designs for Research.* Chicago: Rand McNally, 1966.

6. Cattell, R. B. (ed.). *Handbook of Multivariate Experimental Psychology.* Chicago: Rand McNally and Company, 1966.
7. Cooley, W. W. and P. R. Lohnes. *Multivariate Data Analysis* (New York: John Wiley & Sons, 1971).
8. Cramer, E. M. and R. D. Bock, "Multivariate Analysis," *Review of Educational Research,* 1966, 36 pp. 604–617.
9. Darden, W. R. and R. D. Reynolds, "Male Apparel Innovativeness and Social Context: An Application of Multivariate Analysis of Variance," in *Advances in Consumer Research: Proceedings, Association for Consumer Research* edited by S. Ward and P. Wright (Urbana, Illinois: University of Illinois, 1973), pp. 473–83.
10. Dixon, W. J., editor, *Biomedical Computer Programs.* Los Angeles: Health Science Computing Faculty, School of Medicine, University of California, 1973, pp. 751–64.
11. Finn, J. D. *Multivariance-Univariate and Multivariate Analysis of Variance, Covarience, and Regression: A Fortran IV Program, Version 5* (State University of New York at Buffalo, Department of Educational Psychology, 1972).
12. Green, P. E. and D. S. Tull. *Research for Marketing Decisions* (3rd ed.) Englewood Cliffs: Prentice-Hall, 1970.
13. Hardyck, C. D. and L. F. Petrinovich. *Introduction to Statistics for the Behavioral Sciences,* Second Edition. Philadelphia: W. B. Saunders Company, 1976.
14. Harris, R. J. *A Primer of Multivariate Statistics* (New York: Academic Press, 1975), pp. 231–232.
15. Homans, R. E. and D. J. Messmer, "On the Use of Multivariate Analysis of Variance and Covariance in the Analysis of Marketing Experiments" *Proceedings, American Marketing Association* (Chicago: American Marketing Association, 1976), pp. 398–402.
16. Lindquist, E. F. *Design and Analysis of Experiments in Psychology and Education.* Boston: Houghton Mifflin Company, 1953.
17. Morrison, D. F. *Multivariate Statistical Methods* (New York: McGraw-Hill, 1967).
18. Nelson, R. C. and C. A. Morehouse, "Statistical Procedures Used in Multiple-Group Experiments," *Research Quarterly,* 37, October 1966, p. 44.
19. Perreault, W. D. and W. R. Darden, "Unequal Cell Sizes in Marketing Experiments: Use of the General Linear Hypothesis," *Journal of Marketing Research,* 12 (August 1975), pp. 333–42.
20. Sheth, J. N. *Multivariate Methods for Market and Survey Research* (Chicago: American Marketing Association, 1977), pp. 83–96.
21. Stevens, J. P. "Four Methods of Analyzing Between Variations for the k-Group MANOVA Problem," *Multivariate Behavioral Research* (October, 1972).
22. Tatsuoka, M. M. *Multivariate Analysis: Techniques for Education and Psychological Research.* New York: John Wiley & Sons, 1971.
23. Tukey, J. W. "One Degree of Freedom for Non-Additivity," *Biometrics,* 5, 1949, pp. 232–42.
24. Wilks, S. S., "Certain Generalizations in the Analysis of Variance," *Biometrika,* 24, 1932, pp. 471–94.

25.   Winer, B. J. *Statistical Principles in Experimental Design* (2nd ed.) New York: McGraw-Hill, 1971.
26.   Wind, Yoram and Joseph Denny, "Multivariate Analysis of Variance in Research on the Effectiveness of T.V. Commercials," *Journal of Marketing* Research, 11 (May 1974), pp. 136–42.

SLECTED READINGS

MULTIVARIATE ANALYSIS OF VARIANCE IN RESEARCH ON THE EFFECTIVENESS OF TV COMMERCIALS

YORAM WIND and JOSEPH DENNY*

The effect of TV commercials, or any other form of advertising, can and has been measured on a variety of criteria. Advertising recall, message comprehension, product awareness and knowledge, attitude toward and intentions to buy the advertised product are some of the more frequently used commercial and product-related measures of advertising effectiveness.[1] These measures, being nonpurchase in nature, should ideally be related to the theoretically "optimal" criterion of advertising effectiveness—the present value of the relative profitability of advertising alternatives [5]. Unfortunately, there is no sound theoretical (or even an empirical) basis for selecting any one measure as the single "best" criterion. This may suggest the need for advertising evaluation tests which utilize a number of relevant commercial and product-related criteria. Yet, most current commercial testing procedures, including those that are consistent with the *Dagmar* approach [3], determine the effectiveness of a given commercial based on its performance on a single criterion. It is the purpose of this article to suggest (1) the need for measuring advertising effectiveness on a number of relevant criteria, and (2) a methodology for implementing this approach.

A TYPICAL ON-AIR TV COMMERCIAL TESTING PROCEDURE AND ITS LIMITATIONS

Typical on-air TV commercial testing procedures are based on an experimental design in which a number of test cities are selected for

*Yoram Wind is Professor of Marketing, Wharton School, University of Pennsylvania; Joseph Denny is Marketing Research Director, IT&T Continental Baking Company. The authors wish to acknowledge the helpful suggestions of Paul E. Green for his participation in the design of the orginal application of MANOVA to the on-air TV commercial testing procedure.

[1] For a review of these and other criteria for measuring advertising effectiveness, see [5].

Reprinted by permission of the authors and publisher from the Journal of Marketing Research, *published by the American Marketing Association Vol. xi (May 1974) pp. 136–142.*

TV commercial placement. Test and control groups (about 200 respondents in each) or occasionally only test groups,[2] are selected on a (hopefully) matched basis. Randomly selected respondents are interviewed by telephone according to the following procedure:

1. All respondents are screened for eligibility on demographic and product-class usage characteristics. In addition, they are asked a series of questions concerning their TV viewing frequency and attitude toward the program on which the commercial is to be aired.

2. The control group is asked a series of questions regarding specific attitudes and brand preferences *prior* to watching the program of interest.

3. The test group is asked to watch the program. Subsequent to this, the test group is asked the same kinds of questions presented to the control group regarding specific brand attitudes and preferences. In addition, this group is asked a series of questions regarding the commercial itself, e.g., copy points recalled, general likes and dislikes about the commercial.

With these data, a commercial's effectiveness is determined by the magnitude of the difference between the test and control group with respect to brand preference and specific attitude data. This analysis of the test versus control groups typically consists of a series of "*t*" tests, in which mean ratings on a variety of attitude and preference scales are compared on a univariate basis.

While specific details may vary across various testing procedures, the above steps are typical of this kind of on-air TV commercial testing, and are subject to a number of severe limitations:

1. Despite the use of more than a single response measure (often six to eight), the decision as to which commercial to adopt is often based on a single measure (such as intentions to buy or overall attitude). This is frequently unavoidable since the univariate analysis may lead to inconclusive results, i.e., one commercial may score higher on some measures but lower on others.

2. The response measures—brand-specific attitudes and preferences—are typically intercorrelated themselves.

3. Groups are rarely matched in terms of background variables,

---

[2] When the researcher is concerned only with the relative effectiveness of a given commercial (i.e., with the question of which of a number of commercials is the "best") and is not concerned with measuring the absolute change that a commercial "causes," a simple design which eliminates the control group can be used. In this case, a number of test groups are selected and exposed to the test commercials.

even though data on these variables (demographics, socio-economics and TV viewing, and preference data) are usually obtained during the interviews.

The above problem characteristics serve to cast doubt on the analytical approach of most on-air testing procedures in which the analysis is based on a univariate analysis with no explicit attempt to "match" the background characteristics of the test and control groups.

A PROPOSED MODEL: MULTIVARIATE ANALYSIS OF COVARIANCE

The method proposed to cope with the above problem is multivariate analysis of covariance. This technique—little used by marketing researchers to date—would appear to be particularly appropriate for the type of data described above.

Multivariate analysis of variance [4, 7] is a generalization of the classical ANOVA model to cases in which more than a single criterion variable is involved. In particular, it allows the researcher to make tests of differences involving multiple response variables between two (or more) groups. This technique explicitly takes into account the fact that the two or more response variables—attitudes or preferences—will usually be correlated. That is, significant differences may be found under the univariate approach when, in fact, an overall test of the multiple responses would not show significant differences. Conversely, an overall difference may be detected by this approach in cases where the whole set of univariate tests suggest no differences.

The distinctive feature of the multivariate analysis of variance design is that the dependent variable is a vector-valued variable. This criterion vector is assumed to be multivariate normal with the same within-group dispersion, or variance-covariance matrix, across all groups. This equality of dispersion matrices is the MANOVA extension of the assumption of homogeneity of variance in ANOVA designs. As in the ANOVA design, the research is concerned with testing for the "realness" of the differences among the population centroids, or mean vectors. That is, the research issues concern whether some or all of the populations are centered at different locations in the measurement space spanned by the criterion vector. This situation for the simple MANOVA design for a bivariate variable and the simple ANOVA situation is presented in Figure 1. The discussion in this section is based on [4, pp. 223–41].

To test for equality of group means where the groups have been formed on some a priori "treatment" basis, the ANOVA design utilizes

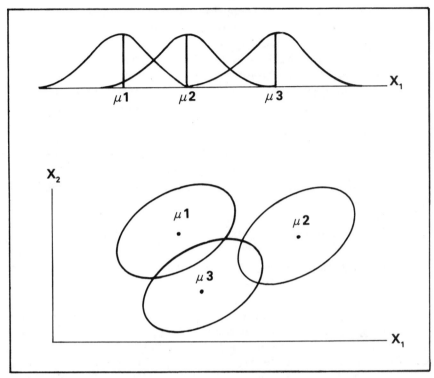

SOURCE: Cooley and Lohnes [4].

Figure 1—A Comparison of the Simple ANOVA and MANOVA Situation

an *F*-ratio (which involves a ratio of the among-group mean square to the within-group mean square adjusted for appropriate degrees of freedom). In the MANOVA case the group centroids are tested for equality using a similar type of ratio which consists of a generalization of the univariate *F*-ratio to a function of the within-groups' and total groups' dispersion matrices [7].

In addition, the covariance feature of the MANOVA model, similar to the covariance feature of ANOVA [1, 2, 7], allows the researcher to adjust statistically response-variable vector components for possible differences in the background variables of the test versus control groups. That is, the researcher is not forced to attempt to match test and control groups on all background-type variables that could affect the responses of interest. Rather, the researcher determines via a regression-type analysis, the effect on the response variables resulting from differences in other independent variables among the test units.

This influence is removed so that the effect of controlled (treatment) variables can be determined independently of the effect of other test unit differences on response.

Finally, most multivariate analysis of variance and covariance programs, such as the BMD-X69 program [6], permit the researcher to make univariate tests in addition to multivariate tests. Thus, if significant overall differences are found, the technique provides ways to examine these differences in more detail.

Multivariate analysis of covariance seems, therefore, ideally suited to the characteristics of on-air TV commercial tests. Moreover, no major changes in data collection procedures are required. Rather the technique provides a superior way to *analyze* the data typically collected in these tests.

AN ILLUSTRATIVE APPLICATION

The advertising research program of a leading manufacturer of frequently purchased household food products included an on-air TV commercial test as the primary vehicle for evaluation and selection of TV commercials. The testing procedure followed the outline presented above for a typical on-air commercial testing procedure. In particular, seven response criteria were used: overall liking, intentions to buy, value for the money, and brand evaluation on four evaluative product attributes. Of these measures, the company used overall attitude and intentions to buy as the two primary criteria, relying on the other measures as secondary criteria. Operationally, however, if a commercial scored significantly higher than the control score on the two primary criteria, it was selected for inclusion in the national TV schedule, regardless of its score on the other five measures. Furthermore, when the two primary criteria led to conflicting results (i.e., a high score on one of them but not on the other) the company tended to rely on a single measure—one of the two primary criteria. The limitations of this procedure, and especially the uncertainty with respect to which response variable to use as the key criterion, led the company to explore the utility of the multivariate analysis of covariance procedure.

The Data

Two samples of 200 respondents each were selected on a matched basis (age range, product, and brand usage). All respondents were interviewed by telephone and asked to provide demographic and product usage data, as well as to watch a specific program on TV. This concluded the first interview with members of the test group. Members

FIGURE 2
THE DATA MATRICES

Control group

"Before" exposure
Background
(Demographics
and product
usage)
1 . . . . k

Brand
evaluation
1 7

"After"
exposure
Exposure to
and evaluation
of program and
commercial
1 . . . . k

Test group

"Before"
exposure
Background
(Demographics
and product
usage)
1 . . . . k

Brand
evaluation
1 7

"After" exposure
Exposure to
and evaluation
of program and
commercial
1 . . . . k

Brand
users
1
. . .
. 100
101

Non-
users
. . .
. 200

Brand
users
1
. . .
. 100
101

Non-
users
. . .
. 200

of the control group, on the other hand, were asked a series of additional questions regarding the brand under investigation. These questions covered the seven response criteria of attitudes toward and intentions to buy the brand.

The second interview with the test group was designed to get their reaction to the brand after exposure to the commercial. They were asked the same kinds of questions presented to the control group in the first interview. In addition, they were asked a series of questions regarding the program and the commercial, e.g., copy points recalled, and general likes and dislikes about the program and commercial. Members of the control group were also interviewed again to check whether they watched the program and to assess their reaction to the program and commercial. This research procedure resulted in two data matrices (for the test and control groups) which are summarized in Figure 2.

Results

The data for the test and control group for both the users and nonusers of the given brand were submitted to the BMD-X69 MANOVA program. The program performs both univariate and multivariate analysis of variance and covariance. We first examine the mean value of each criterion variable, considered separately with and without an adjustment for covariance. An examination of these data (as presented in Table 1) reveals that, whereas overall there are only slight differences between the test and control groups, the main differences lie between the evaluation of the brand by its users and nonusers.

A more detailed examination of the differences between the test and control groups suggests that although most of the differences are in the expected direction—the test group has a higher mean value than the control group—only one of these differences (on intentions to buy) is statistically significant. Thus, the decision of whether to air the commercial or not is inconclusive and depends on management's willingness to rely on the intentions to buy measure and ignore all other measures.

The results of this univariate analysis—the examination of the mean values of each response variable, considered separately—does not take into account the within-cell variation, it ignores the intercorrelations among the seven response variables (summarized in Table 2), and it does not answer the question whether the test and control group are significantly different across the whole set of seven means (one for each criterion variable).

To overcome the limitations, we moved to the multivariate analy-

TABLE 1

UNIVARIATE RESPONSE MEANS OF THE SEVEN CRITERION VARIABLES

	Unadjusted for covariance			Adjusted for covariance		
	Brand users	Non-users	(Marginals)	Brand users	Non-users	(Marginals)
Overall liking						
Test	1.85	2.38	(2.11)	1.85	2.37	(2.11)
Control	1.71	2.60	(2.15)	1.75	2.59	(2.17)
(Marginals)	(1.78)	(2.49)	(2.14)	(1.79)	(2.48)	
Intention to buy						
Test	1.35	2.84	(2.09)	1.35	2.81	(2.08)
Control	1.29	3.13	(2.21)	1.37	3.06	(2.22)
(Marginals)	(1.32)	(2.98)	(2.15)	(1.36)	(2.94)	
Value for the money						
Test	2.09	2.67	(2.38)	2.11	2.65	(2.38)
Control	2.22	2.56	(2.39)	2.22	2.57	(2.39)
(Marginals)	(2.15)	(2.62)	(2.39)	(2.16)	(2.61)	
Product attribute 1						
Test	1.49	2.11	(1.80)	1.51	2.06	(1.78)
Control	1.44	1.91	(1.67)	1.45	1.93	(1.69)
(Marginals)	(1.46)	(2.01)	(1.74)	(1.47)	(1.99)	
Product attribute 2						
Test	1.78	2.25	(2.02)	1.79	2.25	(2.02)
Control	1.64	2.49	(2.06)	1.67	2.46	(2.06)
(Marginals)	(1.71)	(2.37)	(2.04)	(1.73)	(2.35)	
Product attribute 3						
Test	1.67	2.33	(2.00)	1.68	2.32	(2.00)
Control	1.56	2.38	(1.97)	1.61	2.33	(1.97)
(Marginals)	(1.62)	(2.35)	(1.99)	(1.65)	(2.33)	
Product attribute 4						
Test	1.73	2.24	(1.98)	1.72	2.21	(1.97)
Control	1.82	2.29	(2.05)	1.86	2.26	(2.06)
(Marginals)	(1.77)	(2.26)	(2.02)	(1.80)	(2.24)	

Note: The mean values are based on a 5-point Likert-type scale on which 1 is the most preferred (positive) level and 5 the least preferred level. That is, the lower the mean value the more favorable the reaction.

TABLE 2
CORRELATIONS AMONG THE SEVEN RESPONSE MEASURES

	Intention To buy	Overall attitude	Value for money	Attribute 1	Attribute 2	Attribute 3
Test group						
Overall attitude	.43					
Value for money	.25	.39				
Attribute 1	.31	.56	.38			
Attribute 2	.44	.78	.43	.56		
Attribute 3	.47	.77	.49	.57	.86	
Attribute 4	.33	.54	.57	.42	.64	.70
Control group						
Overall attitude	.41					
Value for money	.31	.47				
Attribute 1	.28	.58	.42			
Attribute 2	.40	.68	.48	.53		
Attribute 3	.38	.73	.49	.58	.68	
Attribute 4	.27	.64	.42	.53	.65	.65

sis of variance and covariance which tests whether the whole set of seven means are significantly different between the test and control groups, and between brand users and nonusers. Examination of Table 3 suggests that the test and control groups do not differ significantly with respect to the response centroids, i.e., the commercial had no effect on the set of consumer responses toward the advertised brand. Furthermore, the analysis reveals that there is a significant difference between brand users and nonusers concerning the evaluation of the brand on the seven criterion variables.

Entering eight of the respondent characteristics (age, number of children, income, education, TV watching behavior, liking of the program before and after exposure, and product class consumption) as covariates produced little effect on the analysis. Three of the covariates —amount of TV viewing, liking of program (after) and education—are significantly different, yet their inclusion as covariates exerts no effect on the test vs. control groups, i.e., the null hypothesis of no significant differences in response centroids between the test and control groups is still accepted, after covariate adjustment. Finally, Table 4 presents the result of the univariate ANOVA which examines the differences between the test and control groups and the brand users vs. nonusers on each criterion variable separately. The results suggest that the conclusions drawn from the MANOVA are stable, i.e., there are no signifi-

TABLE 3
SUMMARY OF MULTIVARIATE ANALYSIS OF VARIANCE
AND COVARIANCE (For all 7 criterion variables)

Source of variation	Logarithm of generalized variance[a]	Approximate F statistic[b]
	(d.f. 7; 200)	
Test vs. control (*i*)	33.29	0.69
Brand users vs. nonusers		
(*j*)	33.75	18.33[c]
Interaction (*ij*)	33.30	1.21
Covariates		
TV watching	36.21	2.55[c]
Liking of program (be-		
fore)	34.71	1.25
	36.48	2.79[c]
Product class consump-		
tion	33.95	0.59
Age	35.33	1.79
Education	36.18	2.53[c]
Number of children	34.93	1.44
Income	33.91	0.56

[a] The Logarithm of generalized variance is analogous to a mean square in univariate ANOVA and is based on the computation of a determinant.
[b] The standard computer output of MANOVA includes the U statistic, which involves a ratio of determinants and bears a close relationship to the Wilks's lambda statistic. The BMD X-69 program employs and prints an *F* statistic approximation to the U statistic. (For further discussion of the relationship between the approximate *F* to the U statistic, see [7, 8].)
[c] Significant at the 0.05 level.

cant differences between the test and control and brand users/non-users. The only significant difference is a result of the brand usage.

On the basis of the multivariate analysis, it was recommended not to adopt this commercial. This recommendation was in sharp contrast to the one arrived at by the more traditional analysis which was based on the percentage of respondents in the test group who rated the product (on overall attitude and intentions to buy taken separately) higher than the members of the control group. According to the conventional procedure, the commercial had a +9% on purchase intent (34% for the test vs. 25% for the control group) and +1% on overall attitude (57% for the test vs. 56% for the control group). This inconclusive result led to a split in management—some recommending going ahead with the commercial and some suggesting killing it.

TABLE 4
SUMMARY OF UNIVARIATE ANALYSIS OF COVARIANCE

	Test vs. control	F Statistics Brand users vs. nonusers	Interaction
Overall liking	0.22	27.11[a]	1.52
Intention to buy	0.97	124.76[a]	0.62
Value for the money	0.61	16.73[a]	0.09
Product attribute 1	0.11	20.82[a]	1.54
Product attribute 2	0.05	26.56[a]	0.07
Product attribute 3	0.47	10.93[a]	0.09
Product attribute 4	0.01	9.19[a]	0.46

[a] Significant at the 0.05 level.

CONCLUSIONS

The central premise of this article has been that multivariate analysis of variance and covariance represents a potentially useful approach to marketing experimentation, such as on-air TV commercial testing, involving multiple response variables. The MANOVA approach was utilized in three additional on-air TV commercial tests. Yet, further utilization of this approach in advertising and marketing experimentation would require developments and grounding in three areas [7, Chapter 13]: (1) theoretical understanding of MANOVA and its mathematical foundations; (2) detailed knowledge of the appropriate MANOVA computer algorithms; and (3) examination of the characteristics of various substantive problems in marketing that are relevant for MANOVA analysis.

Some of the more advanced multivariate statistical texts [4, 8], computer program write-ups [6], and advanced marketing research books [7] provide appropriate source material for the first two areas. The third facet of the required developments—exploration of marketing problem areas which can be better solved by the utilization of MANOVA analysis—has not yet been explored. It is the purpose of this section to propose a number of possible areas of application.

In general, multivariate analysis of variance and covariance is a useful analytical design for all situations which require: an ANOVA type model, involving two or more response measures, in which the researcher is interested in evaluating the phenomena under study on the *total set* of response measures.[3] In particular, the MANOVA type design is especially useful for the analysis of a number of advertising and marketing problems.

Advertising Applications

The lack of any theoretical or even empirical basis for selecting a nonpurchase criterion for measuring advertising effectiveness suggests the advantage of measuring advertising effectiveness on a set of multiple criteria. This recommendation is based on the assumption that a person's total response set toward an advertised product or service is a better indicator of his "true" reactions, attitudes, and intended behavior than any single measure. Furthermore, the frequent desire to achieve with advertising more than just a single objective [3] suggests the need to evaluate advertising on multiple criteria.

Given the conceptual advantage of measuring advertising effectiveness on multiple criteria, it is suggested to use multivariate analysis of variance and covariance in all advertising research experiments whether conducted in the "real world" or a laboratory, and whether "pre" or "post" in their orientation. The MANOVA approach is not limited to TV commercial testing and applies to all media (TV, radio, and print) with no major change in the form of application.

Marketing Applications

The conceptual need to measure consumer responses to a marketing variable(s) on more than one criterion is quite evident. Any new marketing strategy or a change in an existing marketing variable (such as package, label, product feature, price, and product positioning) may have a number of effects, ranging from a mere awareness of the change, through some attitude toward the given product and its associated change, to the actual purchase behavior of the product or service. The prevalence of *multiple response* is clearly evident in both marketing practice and literature. An examination of marketing research questionnaires will reveal the widespread practice of measuring the response to any marketing stimulus on a number of diverse questions (responses). Unfortunately, in most cases this has not been accompanied by appropriate analytical procedure for assessing the reactions to the stimulus on all the relevant response measures. The MANOVA approach fills this gap for experimental studies.

---

[3] If the number of response measures is large, the response matrix can be factor analyzed and the MANOVA conducted on the factor scores. Similarly, in some cases of interest one might wish to conduct a factor analysis of covariates and use factor scores in this phase of the analysis.

The advertising and marketing literature has promoted for years the concept of a hierarchy of effects [3]. This model assumes a time sequence between responses (stages in the hierarchy) which enable one to measure a single response at a time. Yet, in practice, most marketing experiments can and do measure more than one of these effects in the same study. It is not uncommon, therefore, to have at a conclusion of an experiment data on awareness, knowledge, liking, preference, conviction, and intended or actual purchase. Given these data, the key question is how to analyze the data in a meaningful way. The intercorrelation of these measures and the conceptual limitations of a univariate analysis suggest the advantages inherent in the utilization of multivariate analysis of variance and covariance.

To this point we have limited the discussion to the measurement of consumer responses to a marketing variable. Consumers, however, are only one of a number of groups aimed at and affected by marketing strategy. An advertising campaign may, for example, be designed not only to generate favorable consumer responses, but also to create favorable trade attitude toward the company and to influence legislators. In this case, one may design an advertising experiment in which the relative effectiveness of an advertising campaign will be measured on the *total set* of responses of all three populations, i.e., the test and control markets will be measured on three sets of responses—those of consumers, trade, and legislators. Again, multivariate analysis of variance and covariance may be utilized to measure the overall effectiveness of the given advertising campaign.

REFERENCES

1. Ackoff, Russell. *The Design of Social Research*. Chicago: The University of Chicago Press, 1953.
2. Cochran, William G. "Analysis of Covariance: Its Nature and Uses," *Biometrics*, 13 (September 1957), 261–81.
3. Colley, Russell H. *Defining Advertising Goals for Measured Advertising Results*. New York: Association of National Advertisers, 1961.
4. Cooley, William W. and Paul R. Lohmes. *Multivariate Data Analysis*. New York: John Wiley & Sons, 1971.
5. Dalbey, Homer M., Irwin Gross, and Yoram Wind. *Advertising Measurement and Decision Making*. Boston: Allyn & Bacon, 1968.
6. Dixon, W. J., ed. *Biomedical Computer Programs*. Los Angeles: Health Science Computing Facility, School of Medicine, University of California, 1965.
7. Green, Paul E. and Donald Tull. *Research for Marketing Decisions*, third edition. Englewood Cliffs, N. J.: Prentice-Hall, in press.
8. Rao, C. R. *Advanced Methods in Biometric Research*. New York: John Wiley & Sons, 1952.

5

Canonical Correlation Analysis

Until recent years canonical correlation analysis was a relatively unknown statistical technique. The availability of canned computer programs has witnessed its increased application to research problems. This chapter introduces the data analyst to the multivariate statistical technique of canonical correlation analysis. Specifically, we shall (1) describe the nature of canonical correlation analysis; (2) illustrate its application; and (3) discuss its potential advantages and limitations. Before reading the chapter you should familiarize yourself with the Definitions of Key Terms.

An understanding of the most important concepts in **canonical correlation analysis** should enable you to:

☐ State the similarities and differences between multiple regression, factor analysis, discriminant analysis, and canonical correlation.

☐ Tell the assumptions that must be met for application of canonical correlation analysis.

☐ State what the canonical root measures, and point out its limitations.

☐ State how many independent canonical functions can be defined between the two sets of original variables.

☐ Compare the advantages and disadvantages of the three methods for interpreting the nature of canonical functions.

☐ Define redundancy and compare it with multiple regression's R^2.

DEFINITIONS OF KEY TERMS

CRITERION VARIABLES.   Dependent variables.

PREDICTOR VARIABLES.   Independent variables.

LINEAR COMPOSITES.   Also referred to as linear combinations, linear compounds, and canonical variates; they represent the weighted sum of two or more variables. In canonical analysis, each canonical function has two separate linear composites (canonical variates), one for the set of criterion variables and one for the set of predictor variables.

CANONICAL LOADINGS.   Referred to by some authors as structure correlations, they measure the simple linear correlation between the independent variables and their respective linear composites and can be interpreted like factor loadings.

CANONICAL CORRELATION.   Measures the strength of the overall relationships between the linear composites of the predictor and criterion sets of variables. In effect, it represents the bivariate correlation between the two linear composites.

CANONICAL ROOTS. Are squared canonical correlations. They provide an estimate of the amount of shared variance between the respective optimally weighted linear composites of criterion and predictor variables.

ORTHOGONAL. A mathematical constraint which specifies that the canonical functions are independent of each other. In other words, the canonical functions are derived so that each function is at a right angle to all other functions when plotted in multivariate space.

REDUNDANCY INDEX. The amount of variance in one set of variables explained by a linear composite of the other set of variables. It can be computed for both the dependent and the independent sets of variables.

5

WHAT IS CANONICAL CORRELATION ANALYSIS?

In Chapter Two, you studied multiple regression analysis, which can be used to predict the value of a single (metric) criterion variable from a linear function of a set of predictor (independent) variables. For some research problems, interest may not center on a single criterion (dependent) variable. Rather, the analyst may be interested in relationships between sets of multiple criterion and multiple predictor variables. Canonical correlation analysis is a multivariate statistical model which facilitates the study of interrelationships among sets of multiple criterion (dependent) variables and multiple predictor (independent) variables. That is, while multiple regression predicts a single dependent variable from a set of multiple independent variables, canonical correlation predicts multiple dependent variables from multiple independent variables.

Hypothetical Example of Canonical Correlation

To further clarify the nature of canonical correlation, let's consider an extension of the example used in the regression chapter. Recall the HATCO survey results focused on using family size and income as predictors of the number of credit cards a family would hold. The problem involved examining the relationship between two independent variables and a single dependent variable.

Suppose HATCO was interested in the broader concept of credit usage by consumers. To measure such a concept it seems logical that HATCO should consider not only the number of credit cards held by the family, but also the average monthly dollar charges of the family on all credit cards. The problem would involve predicting two dependent measures simultaneously (number of credit cards and average dollar charges), and multiple regression is capable of handling only a single dependent variable. Thus, the technique of canonical correlation would be used because it is appropriate where multiple dependent variables are involved.

The problem of predicting credit usage is illustrated in Table 5.1. The two dependent variables used to measure credit usage—number of credit cards held by family and average monthly dollar expenditures on all credit cards—are listed on the left side. The two independent variables selected to predict credit usage—family size and family income—are shown on the right side. By using canonical correlation analysis, HATCO can predict the composite measure of credit usage consisting of both dependent variables, rather than having to

180

TABLE 5.1
PREDICTION OF CREDIT USAGE

Composite of Dependent (Criterion) Variables	Composite of Independent (Predictor) Variables
Number of Credit Cards held by Family Average Monthly Dollar Expenditures on all Credit Cards	Family Size Family Income
Multiple Dependent Variables	} R { Multiple Independent Variables

compute two separate regression equations—one for each of the dependent variables. The result from applying canonical correlation would be a measure of the strength of the relationship between two sets of multiple variables. This measure would be expressed as a canonical correlation coefficient (R) between the two sets.

OBJECTIVES OF CANONICAL ANALYSIS

Canonical correlation analysis is the most generalized member of the family of multivariate statistical techniques (which includes multiple correlation, regression, and discriminant analysis), and is directly related to principal components-type factor analytic models. The goal of canonical correlation is to determine the primary independent dimensions which relate one set of variables to another set of variables. In particular, the objectives may be any or all of the following [6]:

(1) Determining whether two sets of variables (measurements made on the same objects) are independent of one another or conversely, determining the magnitude of the relationships which may exist between the two sets.

(2) Deriving a set of weights for each set of criterion and predictor variables such that the linear combinations themselves are maximally correlated.

(3) Deriving additional linear functions that maximize the remaining correlation, subject to being independent of the preceding set (or sets) of linear compounds.

(4) Explaining the nature of whatever relationships exist between the sets of criterion and predictor variables, generally by measuring the relative contributions of each variable to the canonical functions (relationships) that are extracted.

As noted from the preceding description, canonical analysis is a method for mainly dealing with composite association between sets of multiple criterion and predictor variables. By using this technique, it is possible to develop a number of independent canonical functions that maximize the correlation between the linear composites of sets of criterion and predictor variables.

APPLICATION OF CANONICAL CORRELATION

Discussion of the application of canonical correlation analysis will be organized around two topics. The first will focus on deriving the canonical functions, and the second on the output information.

Deriving the Canonical Functions

The basic input data for canonical correlation analysis is two sets of variables. We assume that each of the two sets of variables can be given some theoretical meaning, at least to the extent that one set could be defined as the independent variable set and the other the dependent variable set. The underlying logic of canonical correlation involves the derivation of a linear combination of variables from each of the two sets of variables so that the correlation between the two linear combinations is maximized.

The application of canonical correlation does not stop with the derivation of a single relationship between the sets of variables. Instead, a number of pairs of linear combinations—referred to as canonical variates—may be derived. The maximum number of canonical variates (functions) that can be extracted from the sets of variables equals the number of variables in the smallest data set, independent or dependent. For example, when the research problem involves five independent (predictor) variables and three dependent (criterion) variables, the maximum number of canonical functions that can be extracted is three.

The derivation of successive canonical variates is similar to the procedure used with unrotated factor analysis. Recall that the first factor extracted accounts for the maximum amount of variance in the set of variables. The second factor is computed so that it accounts for as much as possible of the variance not accounted for by the first factor, and so forth until all factors are extracted. Therefore, successive factors are derived from residual or left-over variance from earlier factors. Canonical correlation analysis follows a similar procedure, but focuses on accounting for the maximum amount of the relationship *between* the two sets of variables rather than within a single set of variables. The result is that the first pair of canonical variates are derived so as to have the highest intercorrelation possible be-tween the two sets of variables. The second pair of canonical variates will then be derived so that they will exhibit the maximum amount of relationship between the two sets of variables which was not accounted for by the first pair of variates. In short, successive pairs of canonical variates are based on residual variance and their respec-tive canonical correlations (which reflect the interrelationship be-tween the variates), become smaller as each additional function is extracted. That is, the first pair of canonical variates will exhibit the highest intercorrelation, the next pair will involve the second largest correlation, and so forth.

One additional point about the derivation of canonical variates. As has been noted, successive pairs of canonical variates are based on residual variance. Therefore, each of the pairs of variates is orthog-onally independent of all other variates derived from the same set of data.

Output Information from Canonical Analysis

The four most important types of output information derived through canonical correlation analysis are: (1) the canonical variates, (2) the canonical correlations between the variates, (3) the statistical significance of the canonical correlations, and (4) the redundancy measure of shared variance for the canonical functions.

Each canonical function consists of a pair of variates, one for each of the subsets of variables entered into the analysis. In other words each canonical function has two variates, one representing the independent variables and the other the dependent variables. The canonical variates are interpreted based upon a set of correlation coefficients, usually referred to as canonical loadings or structure cor-relations.[1] Just as with factor analysis, these coefficients reflect the importance of the original variables in deriving the canonical variates. Thus, the larger the coefficient the more important it is in deriving the canonical variate. Also, criteria for determining the significance of canonical structure correlations are the same as with factor load-ings (see Chapter Six).

Two other types of information provided by a canonical analysis are the canonical correlations and their respective levels of statistical significance. The strength of the relationship between the pairs of variates is reflected by the canonical correlation. When squared, the

---

[1] Some canonical analyses do not compute correlations between the variables and the variates. In such cases the canonical weights would be considered comparable but not equivalent for purposes of our discussion.

canonical correlation represents the amount of variance in one canonical variate that is accounted for by the other canonical variate.
This also may be referred to as the amount of shared variance between the two canonical variates. Squared canonical correlations are
referred to as canonical roots or eigenvalues. As with all correlation
coefficients, canonical or otherwise, various statistics can be utilized
to assess their level of significance—usually expected to be at or
beyond the .05 level to be considered significant.

The last type of information of concern to us at this point is the
redundancy measure of shared variance. At this point an in-depth explanation of this measure could perhaps be confusing to you. Just
remember that using the canonical root as the only measure of shared
variance may lead to some misinterpretation. As a result, a redundancy measure can be computed to provide additional information
concerning the variance shared by the two sets of variables [10].

AN ILLUSTRATIVE EXAMPLE

To illustrate the application of canonical correlation we shall use
variables drawn from the data bank introduced in Chapter One. Recall the data consisted of a series of measures obtained on a sample
of 50 HATCO salespersons. The variables included six social-psychological measures of attitudes, two measures assessing salespersons'
motivation and knowledge, and two different methods of training.

In demonstrating the application of canonical correlation we will
use the first eight variables as input data. The six measures of social-
psychological attitudes (Variables X_1 through X_6) will be designated as
the set of multiple independent variables. The measures of knowledge
and motivation (Variables X_7 and X_8) will be identified as the set of
multiple dependent variables. The statistical problem will involve attempting to predict the knowledge and motivation of the sales
trainees from their social-psychological attitudes.

A canonical correlation analysis was performed using the two
dependent variables and the six independent variables. Two canonical
functions were extracted (note that the number of functions extracted
equals the number of variables in the smallest data set—the dependent). The results are shown in Table 5.2. Four columns of numbers are provided at the top of the table for each of the functions.
The W columns are the canonical weights, the L columns are the
canonical loadings (structure correlations) for each of the original
variables on the canonical variates. The L^2 columns are the squares
of the L column, and the $\%\Sigma L^2$ columns are the results derived by

summing the L^2 columns for the dependent and independent variable sets separately and expressing each variable as a percentage of the total for its respective variate. The lower portion of the table reports the magnitude of the interrelationships between the canonical variates and their respective tests for statistical significance.

The canonical analysis reported in Table 5.2 is based on the Veldman [11] CANONA program. The weights and loadings are a standard output of this package. The L^2 and $\%\Sigma L^2$ were computed by the authors. A similar output is provided by the SPSS package [9].

By now you should be familiar with what canonical correlation analysis is and how it can be applied. Subsequent discussion will focus on the following topics: (1) Which canonical functions should be interpreted?, (2) Which methods should be used for interpreting the nature of canonical function relationships?, and (3) What are the limitations of canonical correlation analysis as a research technique?

TABLE 5.2
HATCO CANONICAL ANALYSIS RELATING
KNOWLEDGE AND MOTIVATION OF SALES
TRAINEES TO SOCIAL-PSYCHOLOGICAL ATTITUDES

Variables	Function I[1]				Function II			
	W	L	L^2	$\%\Sigma L^2$	W	L	L^2	$\%\Sigma L^2$
Dependent Set								
X_7 Knowledge	.67	.91	.83	.49	.71	.41	.17	.55
X_8 Motivation	.74	.93	.87	.51	−.70	−.37	.14	.45
			1.70				.31	
Independent Set								
X_1 Self-Esteem	.18	.76	.58	.33	−.34	.13	.02	.03
X_2 Locus of Control	.06	.01	.00	.00	−.17	.17	.03	.05
X_3 Alienation	.69	.64	.41	.23	.04	.11	.01	.02
X_4 Social Responsibility	.39	.38	.14	.08	−.67	−.68	.46	.73
X_5 Machiavellianism	.58	.74	.55	.31	.50	.26	.07	.11
X_6 Political Opinion	−.04	.30	.09	.05	.39	−.20	.04	.06
			1.77				.63	
Canonical R		.94				.48		
Canonical Root		.88				.23		
Chi Square Value		95.51				11.92		
Degrees Freedom		7.00				5.00		
Level of Significance		0.00				0.04		

[1] The numbers in the W columns are the canonical weights; the numbers in the L columns are the canonical loadings; the L columns are squared and reported in the L^2 columns; and the percentage sums of loadings are shown in the $\%\Sigma L^2$ columns.

Which Canonical Functions Should Be Interpreted?

As with research using other statistical techniques, the most common practice is to analyze those functions whose canonical correlation coefficients are statistically significant beyond some level, typically .05 or above. Thus, variables in each set which contribute heavily to shared variance for these functions are considered to be related to each other. The other independent functions are deemed insignificant, and the relationships among the variables are not interpreted [1].

The authors believe that the use of a single criteria such as the level of significance is too superficial. Instead, it is recommended that three criteria be used in conjunction with each other as a means of deciding which canonical functions should be interpreted. By interpretation we mean examining the canonical loadings to determine how the original variables from the two data sets are related. The three criteria are: (1) the level of statistical significance of the function, (2) the magnitude of the canonical correlation, and (3) the redundancy measure for the percentage of variance accounted for from the two data sets.

Level of Significance. The level of significance of a canonical correlation which is generally considered to be the minimum for interpretation is the .05 level. The .05 level (along with the .01 level) has become the generally accepted level for considering a correlation coefficient statistically significant. This is largely due to the availability of tables for these levels, borrowed from other disciplines where higher confidence levels are desired. These levels are not impervious however, and researchers from various disciplines frequently must rely on results based on lower levels of significance.

Several statistics can be used for evaluating the significance of canonical roots. The most widely used test and the one normally printed-out in canned computer programs is Bartlett's approximate Chi-square [3].

Magnitude of the Canonical Relationships. The size of the canonical correlations also should be considered in deciding which functions to interpret. No generally accepted guidelines have been established regarding acceptable sizes for canonical correlations. Rather, the decision is usually made based on the contribution of the findings toward better understanding of the research problem being studied. It seems logical that the guidelines suggested for significant factor loadings (see Chapter Six) might be useful with canonical correlations. This is particularly true when one considers the fact that canonical correlations refer to the variance explained in the canonical variates (linear composites) and not in the original variables.

Redundancy Measure of Shared Variance. Recall that squared canonical correlations (Roots) provide an estimate of the shared variance between the canonical variates. While this is a simple and appealing measure of the shared variance, it may lead to some misinterpretation. This is because the squared canonical correlations represent the variance shared by the linear composites of the sets of criterion and predictor variables, and not the variance extracted from the sets of variables [1]. Thus, a relatively strong canonical correlation may be obtained between two linear composites (canonical variates), even though these linear composites may not extract significant portions of variance from their respective sets of variables [10].

Since canonical correlations may be obtained which are considerably larger than previously reported bivariate and multiple correlation coefficients, there may be a temptation to assume that canonical analysis has uncovered substantial relationships of conceptual and practical significance. Before such conclusions are warranted, however, further analysis involving measures other than canonical correlations must be undertaken to determine the amount of the dependent variable variance that is accounted for or shared with the independent variables [7].

To overcome the inherent bias and uncertainty in using canonical roots (squared canonical correlations) as a measure of shared variance, a redundancy index has been proposed [10]. The redundancy index is the equivalent of computing the squared multiple correlation coefficient between the total predictor set and each variable in the criterion set, and then averaging these squared coefficients to arrive at an average R^2. It provides a summary measure of the ability of a set of predictor variables (taken as a set) to explain variation in the criterion variables (taken one at a time). As such, the redundancy measure is perfectly analogous to multiple regression's R^2 statistic, and its value as an index is similar [4].

The redundancy index of the shared variance of the HATCO canonical functions (see Table 5.2) is illustrated in Table 5.3. Results are shown for both the dependent and independent sets, although in most instances the researcher is only concerned with the variance extracted from the dependent variable set. The redundancy index for the example problem indicates that 78 percent of the variance in the dependent variables has been explained by the canonical variate for the independent variable set. While the redundancy index lowers the

TABLE 5.3
REDUNDANCY INDEX FOR HATCO CANONICAL FUNCTIONS

Canonical Function	Root	Variance Extracted	Redundancy	Proportion Of Total Redundancy
		Dependent Set		
	R^2	VED^1	$R^2 \times VED$	
1	.88	.85	.75	.96
2	.23	.15	.03	.04
		1.00	.78	1.00
		Independent Set		
	R^2	VEI^1	$R^2 \times VEI$	
1	.88	.30	.26	.90
2	.23	.11	.03	.10
		.41	.29	1.00

[1] VED refers to the variance extracted from the dependent set of variables. It is computed by summing the L^2 column for the dependent variables (see Table 5.2) and dividing the result by the number of dependent variables. VEI refers to the variance extracted from the independent variables. It is computed in the same manner as VED except the corresponding values for the independent variables are used. A general formula for variance extracted is:

$$VE = \frac{\Sigma L^2}{\# \text{ variables}}$$

shared variance estimate obtained from the canonical roots, it still represents a substantial amount of shared variance. Moreover, it provides the analyst with a much more realistic measure of the predictive ability of canonical relationships.

The question arises as to what is the minimum acceptable redundancy index to justify interpretation of canonical functions? Just as with canonical correlations, no generally accepted guidelines have been established. The analyst would have to judge each canonical function in light of its theoretical and practical significance to the research problem being investigated to determine if the redundancy index is sufficient to justify interpretation. Also, a test for the significance of the redundancy index has been developed [2], although it has not been widely utilized.

Interpretation Methods for Canonical Functions

If the canonical relationship is statistically significant and the magnitude of the canonical root and the redundancy index is acceptable, then the analyst would still like to make substantive interpretations of the results. This involves examining the canonical functions to determine the relative importance of each of the original variables in deriving the canonical relationships. Three methods have been proposed: (1) canonical weights, (2) canonical loadings (structure correlations), and (3) canonical cross loadings.

Canonical weights. The traditional approach to interpreting canonical functions involves examining the sign and magnitude of the canonical weight assigned to each variable in computing the canonical functions. Variables with relatively larger weights contribute more to the functions and vice versa. Similarly, variables whose weights have opposite signs would exhibit an inverse relationship with each other and those with the same sign a direct relationship. However, interpreting the relative importance or contribution of a variable by its canonical weight is subject to the same criticisms associated with the interpretation of beta weights in regression techniques. For example, a small weight may either mean its corresponding variable is irrelevant in determining a relationship, or that it has been partialed out of the relationship because of a high degree of multicollinearity. Another problem with the use of canonical weights is that they are subject to considerable instability (variability) from one sample to another. This is because the computational procedure for canonical analysis yields weights that maximize canonical correlations for a particular sample of observed dependent and independent variable sets [7]. These problems suggest considerable caution in using canonical weights to interpret the results of canonical analysis.

Canonical loadings. In recent years canonical loadings have been increasingly used as a basis for interpretation because of the deficiencies in utilizing weights. Canonical loadings, referred to sometimes as structure correlations, measure the simple linear correlation between an original observed variable in the dependent or independent set and the set's canonical variate. The methodology considers each independent canonical function separately and computes the within-set variable-variate correlation [1]. That is, for each set of variables, dependent and independent, the correlation is computed between each original observed variable and its respective canonical variate. Thus, the canonical loading reflects the variance which the observed variable shares with the canonical variate, and can be interpreted like a factor loading in assessing the relative contribution of each variable to each canonical function.

Canonical loadings like weights may be subject to considerable variability from one sample to another. This variability suggests that loadings and hence the relationships ascribed to them are sample specific, due to change or the result of extraneous factors [7]. Canonical loadings are considered relatively more valid than weights as a means of interpreting the nature of canonical relationships. But the analyst still must be cautious when using loadings for interpreting canonical relationships, particularly with regard to the external validity of the findings.

   Canonical Cross-Loadings. The computation of canonical cross-loadings has been suggested as an alternative to conventional loadings [4]. This procedure involves correlating each of the original observed dependent variables directly with the independent canonical variate. Recall that conventional loadings correlate the original observed variables with their respective variates after the two canonical variates (dependent and independent) are maximally correlated with each other. Thus, cross-loadings do provide a more direct measure of the dependent-independent variable relationships by eliminating an intermediate step involved in conventional loadings.

How To Interpret Canonical Functions

   Several different methods for interpreting the nature of canonical relationships were discussed in the last section. The question remains, however, which method should the analyst use? Since most canonical analysis problems necessitate the use of a computer, the analyst frequently must use whichever method is available. The use of cross-loadings is the preferable approach, but none of the popular canned computer packages provide cross-loadings as a standard output. Thus, the analyst either must compute these or select another method. The widely available SPSS package [9] and the lesser known Veldman [11] package provide canonical loadings while the BMD [5] package outputs only the canonical weights. The canonical loadings approach is somewhat more valid than the use of weights, therefore whenever possible it is recommended that the loadings approach be utilized.

   To illustrate how a canonical function might be interpreted, let's use the canonical loadings approach and refer back to the data reported in Table 5.2. In interpreting canonical relationships, the analyst should examine each canonical function separately and look for clusters of variables on both sides of the function which exhibit high loadings on their respective variates. For example, looking at function one it can be noted that both dependent variables have high loadings—.91 and .93 for variables X_7 and X_8, respectively. At the same time, the independent set variables X_1, X_3, and X_5 are all closely related to both of the dependent variables. Another way of stating this is that these five variables contribute relatively larger amounts to the shared variance expressed by the canonical root than do the three independent variables with lower loadings (X_2, X_4, and X_6).

The interpretation of the second canonical function involves additional considerations. Whereas with the first function several variables exhibited relatively high loadings (.64 or above), in the second function most of the loadings are relatively low. Thus, the question arises as to whether one should consider any of the relationships significant. Each canonical function must be evaluated on its own merits, but in general we would expect the loadings to be relatively smaller on successive functions. Therefore, different minimum loadings can be used on different functions. As pointed out earlier, the criteria for significant factor loadings also should be used with canonical loadings. Using these guidelines (see Chapter Six), we could conclude for function two that both dependent variables are significantly related to only one independent variable—X_4.

A question might also arise as to why variable X_8 with a loading of .37 on function two was considered significant, whereas variable X_4 with a loading of .38 was not considered significant on the first function? The answer is that variable X_4 and even variable X_6 with a loading of .30 on the first function could have been considered significant. But they were not because the analyst should focus on the most significant relationships. Such a procedure will facilitate interpretation of the findings and make them conceptually more manageable. The decision rule for not considering variables X_4 and X_6 significant was that several variables had high loadings, and the difference between these loadings and the higher ones was considerable (.64 for X_3 minus .38 for $X_4 = .26$).

A final point for discussion of interpretation is how do we use the signs? The answer is that like-signs indicate a direct relationship and opposite-signs an inverse one. For example, on function one all the signs are positive. Therefore all the variables are directly related. But on function two variables X_4 and X_8 are negative and X_7 is positive. Thus, variables X_4 and X_7 are inversely related, and X_4 and X_8 are directly related.

Limitations of Canonical Analysis

Applications of canonical correlation analysis have expanded substantially in recent years. Like other multivariate techniques, however, it is not without certain limitations. When selecting a statistical technique for data analysis and interpreting the results of canonical correlation, it is important to keep these limitations in mind. The following limitations should be closely observed when applying canonical correlation: (1) the canonical correlation reflects the variance shared by the linear composites of the sets of variables, and not the variance extracted from the variables; (2) canonical weights derived in computing canonical functions are subject to a great deal of instability; (3) canonical weights are derived to maximize the correlation between linear composites, not to maximize the variance extracted, and (4) it is difficult to identify meaningful relationships between the subsets of independent and dependent variables because precise statistics have not yet been developed to interpret canonical analysis and we must rely on inadequate measures such as loadings or cross-loadings [7]. These limitations are not meant to discourage the use of canonical correlation. Rather they are pointed out so that the effectiveness of canonical correlation as a research tool will be enhanced.

SUMMARY

Canonical correlation analysis is a highly useful, powerful technique for exploring the relationships among multiple criterion and multiple predictor variables. The technique is primarily descriptive, although it may be used for predictive purposes. Results obtained from a canonical analysis should suggest answers to questions concerning the number of ways in which the two sets of multiple variables are related, the strengths of the relationships, and the nature of the relationships so defined.

Canonical analysis enables the data analyst to combine into a composite measure what otherwise might be an unmanageably large number of bivariate correlations between sets of variables. It is useful for identifying overall relationships between multiple independent and dependent variables, particularly when the data analyst has little a priori knowledge about relationships among the sets of variables.

END OF CHAPTER QUESTIONS

1. Under what circumstances would you select canonical correlation analysis instead of multiple regression as the appropriate statistical technique?
2. What are the three criteria you should use in deciding which canonical functions should be interpreted? Explain the role of each.
3. Explain how you would go about interpreting the nature of a canonical correlation analysis.
4. What is the relationship between the canonical root, the redundancy index, and multiple regression's R_2?

5. What are the limitations associated with the technique of canonical correlation analysis?

REFERENCES

1. Alpert, Mark I., and Robert A. Peterson, "On the Interpretation Canonical Analysis." *Journal of Marketing Research,* Vol. IX, (May, 1972), p. 187.
2. Alpert, Mark I., Robert A. Peterson, and Warren S. Martin. "Testing the Significance of Canonical Correlations," *Proceedings,* American Marketing Association, E. M. Mazze (ed.), Volume 37, 1975, pp. 117-119.
3. Bartlett, M. S. "The Statistical Significance of Canonical Correlations." *Biometrika,* Vol. 32, 1941, p. 29.
4. Cooley, William W. and Paul R. Lohnes, *Multivariate Data Analysis.* New York: John Wiley and Sons, Inc., 1971.
5. Dixon, W. J. *Biomedical Computer Programs,* UCLA, 1974.
6. Green, Paul, and Donald Tull, *Research for Marketing Decisions.* Englewood Cliffs, New Jersey: Prentice-Hall, Inc., 1975, 3rd Edition.
7. Lambert, Z., and R. Durand, "Some Precautions in Using Canonical Analysis." *Journal of Marketing Research* Vol. XII (November, 1975), p. 468.
8. Meredith, W., "Canonical Correlations with Fallible Data." *Psychometrika,* Vol. 29 (1964), pp. 55-56.
9. Nie, N., C. Hull, J. Jenkins, K. Steinbrenner, and D. Bent, *Statistical Package for the Social Sciences,* New York: McGraw-Hill, 1975.
10. Stewart, Douglas, and William Love, "A General Canonical Correlation Index." *Psychological Bulletin,* Vol. 70 (1968), pp. 160-63.
11. Veldman, Donald. *Fortran Programming for the Behavioral Sciences.* New York: Holt, Rinehart and Winston, Inc., 1967.

SELECTED READINGS

A MULTIVARIATE ANALYSIS OF PERSONALITY AND PRODUCT USE

DAVID L. SPARKS and W. T. TUCKER*

Introduction

Despite the general failure of empirical studies over the past ten or more years to locate important relationships between personality and consumptive behavior, there remains among students of

*David L. Sparks is Assistant Professor of Marketing, University of Richmond and W. T. Tucker is Professor of Marketing Administration, University of Texas at Austin. The authors express their appreciation to Professor Grady D. Bruce for his valuable suggestions during the course of this study.

Reprinted by permission of the authors and publisher from the Journal of Marketing Research, *published by the American Marketing Association Vol. ix (May 1972) pp. 187–92.*

marketing this item of faith: behavior in the marketplace is critically reflective of individual personality. The corollary of that belief is that the measuring instruments or statistical techniques (or both) that have been commonly used in empirical work are incapable of giving more than glimpses of the structures and processes involved.

Statistical Techniques

Most of the work attempting to relate personality to consumer behavior has used bivariate inferential techniques or regression including multiple correlation. This implies the view (probably not held by any researcher) of personality as a bundle of discrete and independent traits which either do not interact or do so only in the simple sense that a number of diverse forces can be resolved into a *single* vector.

A recent study by Kernan notes that canonical analysis, alone or in conjunction with hierarchical grouping, can suggest the existence of molar personality types that are essentially synthesized out of the individual traits of a simple personality test [6]. Since Kernan's data delivered only one significant canonical root (at the .10 level), he could not use that technique to draw inferences about the complexity of personality trait interaction; but a hierarchical clustering of subjects based on choice strategies in a game playing situation posited four synthetic character types in which total personalities rather than specific traits seemed to be the operant variables.

The present study parallels the Kernan research with the intention of using hierarchical grouping in the same way, unless canonical analysis infers several significant roots.

In effect, the statistical techniques used in many previous studies relating personality to consumer behavior [2, 3, 7, 8, 9, 10] probably constitute a part of the "inadequate theoretical framework" referred to by Brody and Cunningham [1].

Measuring Instruments

Psychologists are no more elated than those in marketing with current personality theory or the attendant measuring instruments. There is no persuasive theoretical basis for preferring one sort of personality test to another, despite a great variety of tests. On one hand, instruments like the Edwards or the California Personality Inventory measure a host of individual traits; on the other, the I-O scale locates everyone at some point on a unidimensional continuum. (Clinical techniques requiring subjective judgments are disregarded here for operational reasons.)

Additionally, instruments may be roughly categorized into two subclasses, those asking largely for: (1) direct reports on thoughts and feelings, and (2) reports of activities, actual or preferred. Preferences for one or the other of these subclasses will in some measure depend upon the way the experimenter regards personality. It is legitimate to think of personality as an intimate aspect of the cognitive and affective organization of the central nervous system (or the total organism). It is equally legitimate to regard it as a verbal construct describing behavioral regularities. When someone is described as anti-intraceptive or rigid, it is the cognitive and affective organization that is the principal reference. To call someone sociable or kleptomaniac is to classify him behaviorally with little regard for central processes. This dichotomy is not rigorous; a number of personality tests, Cohen's CAD for instance [2], ask for cognitive evaluations of behavior or otherwise provide mixed cases.

RESEARCH DESIGN

A sample of 190 college students (173 of whom accurately completed forms) chosen for their availability in introductory marketing classes were used to explore the relationship between consumer behavior and personality. The choice of such a sample (in this case all of the males present in particular classes on a particular day) seems appropriate when the effort is to locate the existence of relationships rather than to describe or define them for particular universes. Beyond this, the sample method was essentially that of Kernan [6] and Tucker and Painter [9]. Both previous studies showed that the frequency distributions on the Gordon Personal Profile [5] and Gordon Personal Inventory [4] for such a sample varied little from those of groups on which the test was normed.

The use of the Gordon tests was based on several considerations: (1) the bias of the authors toward the behaviorally-oriented rather than the cognitively-oriented test as relevant to consumer behavior, (2) the previous and partially successful use of that test [6, 9], and (3) the short time required for subjects to complete the tests. The fact that eight traits isolated by the test are not fully independent is of concern but seems of less consequence than the test's demonstrated ability to differentiate people with regard to the kinds of behavior under study.

The instrument to measure the subjects' product use had 17 multiple-choice questions. The products, considered to be typical for this subject group, were: headache remedies, mouthwash, men's

TABLE 1
CORRELATION MATRIX: PRODUCT USE AND PERSONALITY TRAIT

Product	Ascendancy	Responsibility	Emotional stability	Sociability	Cautiousness	Original thinking	Personal relations	Vigor
Headache remedy	.0254	−.1391	−.2104[a]	.1490	−.0073	−.0649	−.0875	−.0907
Mouthwash	.0702	−.0983	−.1308	.1125	.1501[a]	−.0242	.0443	−.1238
Men's cologne	.1473	−.1066	−.1222	.2599[a]	.1247	.0715	−.0459	.0008
Hair spray	−.0580	−.1241	−.0725	.0388	−.0824	−.0668	−.0664	−.0159
Shampoo	.1735[a]	−.1420	.0729	.1459	−.0449	.0757	.0412	.0116
Antacid remedy	.0217	−.1521[a]	−.2692[a]	.0393	−.1222	−.0974	−.1119	−.0886
Playboy	.1293	−.0218	.0787	.2621[a]	−.1038	.0650	.0169	.1185
Alcoholic beverages	.2001[a]	−.1605[a]	.0159	.1973[a]	−.2861[a]	.0041	−.1436	.0261
Brush teeth	−.1324	−.0418	.0196	−.0624	−.0663	.1645[a]	.0329	−.1074
Fashion adoption	.2892[a]	−.1647[a]	−.0628	.3858[a]	−.0919	.0924	.0838	.0557
Complexion aids	.0065	−.0591	−.0106	.0845	−.1131	−.0826	−.0667	−.0902
Vitamin capsules	.1384	−.1197	−.1759[a]	.1288	−.0855	.0963	−.0414	.0016
Haircut	−.0587	.0616	.0655	−.0774	−.0670	−.0247	−.0394	−.0311
Cigarettes	.0869	−.1465	−.1213	.0954	−.1313	.1408	−.0376	−.0305
Coffee	−.0413	−.0265	−.1478	−.0185	.0403	−.0781	−.0683	−.0734
Chewing gum	.1645[a]	−.1035	−.1165	.2581[a]	−.1209	−.0447	.0433	−.0446
After-shave lotion	.0506	.1016	.0429	.0751	.0091	.1288	.0168	.0676

[a] Indicates correlation coefficient is significant at the .05 level.

cologne, hair spray, shampoo, antacid remedies, *Playboy,* alcoholic beverages, complexion aids, vitamins, cigarettes, coffee, chewing gum, and after-shave lotion. In addition, subjects were asked how often they brushed their teeth and had their hair cut. Another question asked about their adoption of new clothing fashions. Response categories were, generally: never, less than once a week, about once a week, more than once a week but less than once a day, and about once a day. For five of the products, dichotomous or specially worded response categories were required. While these products were not a complete inventory of typical products, they did represent a reasonable number and were considered sufficient for this investigation.

A pretest of the 17-item product-use questionnaire with 62 male undergraduates led to minor changes in the question wording and response categories. A varimax factor analysis of the pretest data showed the 17 questions to be almost completely independent, the last of the 17 factors extracting nearly as much variance as the first. While desirable in one sense, independence as extensive as this raises critical issues which will be discussed later.

FINDINGS

A correlation analysis of the data (Table 1) shows essentially the the same weak and spotty relationships between personality traits and particular product use reported previously. It may lead one to conclude that some two percent of the variance in the use of mouthwash may be accounted for by cautiousness or that some six percent of the variance in the use of men's cologne is associated with sociability. The total of 18 significant but low correlations in a matrix where seven would be expected to occur by chance may be persuasive that something is responsible, but the findings seem to be of minimal value.

Canonical analysis provides both a more persuasive case for the relationship under study and some hints concerning the kinds of personality structures involved. That is far from saying that canonical analysis illuminates the field; it is notoriously difficult to interpret beyond the significance levels of R's associated with particular roots.

Table 2 shows the first three canonical roots with R's of .606, .548, and .413. These have significance levels of .0001, .0002, and .0752 respectively, leaving little doubt that there are significant relationships involved. More interesting, since the basic relationship involved has not really been in doubt, is the nature of the relationship

TABLE 2
RESULTS OF THE CANONICAL ANALYSIS

Variables	Canonical coefficients		
	1	2	3
Criterion set (product use)			
Headache remedy	−.0081	−.4433	.1123
Mouthwash	−.1598	−.4538	.2809
Men's cologne	.2231	−.1935	−.2121
Hair spray	.0664	.0706	.0857
Shampoo	.3784	.1587	−.0063
Antacid remedy	−.1421	−.1746	−.3226
Playboy	.1511	.1591	.5220
Alcoholic beverages	.4639	.3098	−.1329
Brush teeth	−.1879	−.0152	.2341
Fashion adoption	.3226	−.3993	.0856
Complexion aids	−.0243	.0925	.1799
Vitamin capsules	.2870	−.0599	−.4975
Haircut	−.1698	.1855	−.0170
Cigarettes	.4065	.0551	−.2894
Coffee	−.2441	−.2453	.1330
Chewing gum	.2051	−.1320	.1342
After-shave lotion	−.0270	.3022	.0108
Predictor set (personality traits)			
Ascendancy	.0182	−.0517	−.4375
Responsibility	−.5125	.0777	−.1688
Emotional stability	.4309	.6405	.4880
Sociability	.6072	−.3597	.6199
Cautiousness	−.2869	−.5959	.2438
Original thinking	.2377	.1620	−.3076
Personal relations	−.1245	−.0567	.0369
Vigor	.1681	.2592	.0481
Roots	.3671	.3000	.1711
Canonical R	.606	.548	.413
x^2	72.7419	56.7026	29.8417
d.f.	24	22	20
Probability	.0000	.0002	.0752

suggested. The meanings of the roots can be crudely approximated by extracting the items with heavy loadings from the predictor and criterion sets, somewhat simplifying the picture. In this case, items with coefficients above .30 are used.

The first root is associated with the use of shampoo, alcoholic beverages, cigarettes, and early fashion adoption. Those involved are best described as sociable, emotionally stable, and irresponsible (minus responsibility). The relationships are intuitively acceptable, although they are certainly not the only ones that would be so. Nevertheless, it makes sense to think that early fashion adopters are those particular sociables who are also emotionally stable (not easily upset) and also somewhat irresponsible (responsibility has previously been associated with modal behavior [9]).

The second root is associated with (again converting signs verbally for ease of expression) the use of headache remedies and mouthwash, late fashion adoption, and infrequent use of after-shave lotion. The personality characteristics are sociability, cautiousness, and emotional instability. At this point there emerges a clear advantage to the methodology. Both early and late fashion adoption are related to sociability, but in different personality contexts.

This seems to be exactly the kind of relationship personality theory implies: not a simple connection between sociability and

TABLE 3
RESULTS OF CLUSTER ANALYSIS

	Cluster					
	1	2	3	4	5	6
Personality trait:						
Kernan's study[a]						
Ascendancy	81.0	19.0	81.0	43.0		
Responsibility	43.0	11.0	19.0	80.0		
Emotional stability	40.0	15.0	40.0	77.0		
Sociability	77.0	12.0	77.0	39.0		
Cautiousness	54.0	31.0	5.0	69.0		
Original thinking	72.0	31.0	45.0	43.0		
Personal relations	70.0	10.0	54.0	57.0		
Vigor	81.0	13.0	38.0	63.0		
Personality trait: present study[a]						
Ascendancy	66.6	54.9	50.8	52.3	49.4	63.0
Responsibility	44.2	55.2	49.2	47.6	61.7	41.1
Emotional stability	43.7	57.4	54.3	50.7	47.8	44.4
Sociability	63.6	49.1	42.9	37.9	36.4	59.8
Cautiousness	40.2	45.1	46.6	51.5	62.2	33.5
Original thinking	52.2	54.5	53.1	48.3	43.9	51.0
Personal relations	49.4	42.3	45.5	48.7	40.3	39.3
Vigor	49.4	61.2	61.5	54.1	44.3	52.3

[a] Mean percentile scores.

early fashion adoption, but a more complex one in which sociability combined with emotional stability and irresponsibility is oriented toward one sort of action while sociability combined with emotional instability and cautiousness is oriented toward its opposite.

In the third root (with the marginal significance level of .075) sociability again characterizes the individual, but in this context the relationship with fashion adoption is very low and there is an association with light or no use of cigarettes, again a reversal of the variate-to-variate relationship suggested by the first root.

The most obvious explanation for these findings lies in the notion that it is the person in some gestalt in which the entire personality and the entire situation form a particular configuration, who acts, not the individual personality trait. But this view includes the possibility that the most useful approach to the subject is to measure individual personality characteristics and synthesize the molar personality from such measures. The relationships of the above canonical analysis suggest that even a simple model based on trait interaction could prove more predictive than a trait-by-trait approach. Nevertheless, some of the relationships suggested could stem in large part from nonlinearity. Further, canonical analysis is a linear technique which can only indirectly suggest the presence of certain possible nonlinear associations while leaving others occult.

The present study parallels that of Kernan [6] closely enough that there is some interest in seeing whether a hierarchical clustering of subjects on the basis of their reported product-use behavior approximates the interesting personality profiles that related to particular game playing strategies. Table 3 shows the four clusters Kernan located and the six clusters that seem to best describe the present data. No persuasive case can be made for similarities in grouping, although the imaginative mind can perceive parallels. Nor does the cluster analysis, when compared with product use, add to the conclusions available through canonical analysis alone in this case. It seems possible that the near-fantasy situation of game playing in relative isolation may give freer play to personality expression than consumption patterns which operate under social, economic, and habitual constraints.

The annoying fragmentation of 17 questions into 17 factors of approximately equal magnitude is not readily explained. It is difficult to conceive that the frequency of use of mouthwash, men's cologne, hair spray, shampoo, and after-shave lotion are essentially independent behaviors not tied together. The problem may lie in the methodology, although it is difficult to understand how the response categories could mask associations when most were used by fairly large numbers of subjects. Yet on both pretest and test the same lack of structure appeared. The kind of post hoc explanations that come to mind do little to reassure one that there are not large areas of dissociated events in consumer behavior that will require explanatory models far more complex or far more numerous than one would wish.

CONCLUSIONS

The association of identical personality traits (within different sets of personality traits) with diverse consumer behavior suggests that trait interactions or nonlinear relationships may compose a significant portion of the personality-behavior relation. This may partially explain the difficulty in empirically demonstrating the commonly accepted hypothesis that personality influences consumer activities. Inferential techniques do not generally lend themselves to the location of the sorts of relationships implied by these findings.

The apparent lack of correlation among product-use patterns suggested by factor analyses of questionnaire responses leads to the conclusion that a general model applicable to all consumer behavior would prove extremely complex. The alternative of exploring personality in connection with particular behavior or particular products seems therefore the only current application to practical marketing problems.

The particular relationships among traits suggested by this study should be considered as merely representative of the sorts of interrelations that can occur. In all probability a study of other subjects, and other products, or other sorts of behavorial differences would show the relevance of different trait combinations.

REFERENCES

1. Brody, Robert P. and Scott M. Cunningham. "Personality Variables and the Consumer Decision Process," *Journal of Marketing Research*, 5 (February 1968), 50–7.
2. Cohen, Joel B. "An Interpersonal Orientation to the Study of Consumer Behavior," *Journal of Marketing Research*, 4 (August 1967), 270–8.
3. Evans, Franklin B. "Psychological and Objective Factors in the Prediction of Brand Choice: Ford Versus Chevrolet," *Journal of Business*, 32 (October 1959), 340–69.
4. Gordon, Leonard V. *Gordon Personal Inventory*. New York: Harcourt, Brace, & World, 1963.
5. ————. *Gordon Personal Profile*. New York: Harcourt, Brace, & World, 1963.

6. Kernan, Jerome B. "Choice Criteria, Decision Behavior, and Personality," *Journal of Marketing Research*, 5 (May 1968), 155–64.
7. Koponen, Arthur. "Personality Characteristics of Purchasers," *Journal of Advertising Research*, 1 (September 1960), 6–12.
8. Pessemier, Edgar A., Philip C. Burger, and Douglas J. Tigert. "Can New Product Buyers be Identified?" *Journal of Marketing Research*, 4 (November 1967), 349–55.
9. Tucker, W. T. and John J. Painter. "Personality and Product Use," *Journal of Applied Psychology*, 45 (October 1961), 325–9.
10. Westfall, Ralph. "Psychology Factors in Predicting Product Choice," *Journal of Marketing*, 26 (April 1962), 34–40.

ON THE INTERPRETATION OF CANONICAL ANALYSIS

MARK I. ALPERT and ROBERT A. PETERSON*

More than 35 years ago, Hotelling derived a method to deal with the general problems of relating two sets of variables measured across a group [11, 12]. This technique, canonical analysis or the method of canonical correlations, has recently received increased attention, and efficient computational procedures and several proofs and extensions have been offered, e.g., [1, 5, 6, 9, 13, 14, 15, 17, 19, 20, 23, 24].[1] Simultaneously, computation of canonical correlations has been greatly aided by the widespread availability of high-speed computing machinery and related software packages [4, 22, 27].

Unfortunately, methods of interpreting canonical correlations have lagged behind the technical advances [13]. The objectives of this article are to explain alternative interpretations and practical (as opposed to purely statistical) guides, and to help avoid problems which may be obscured by the easy availability of "canned" computer output.

In general, the objectives of canonical analysis are:
1. To determine vectors of weights for each set of variables such that linear combinations of the respective variables are maximally correlated. This goal implies optimal prediction of linear combinations of variables (variates) from one vector, given variable values in the other vector.

    * Mark I. Alpert and Robert A. Peterson are now promoted to Professors of Marketing Administration, The University of Texas at Austin.
    [1] For simplicity of exposition, only the two-set case will be discussed here, although it is possible to extend it to the *m*-set case [9].
Reprinted by permission of the authors and publisher from the Journal of Marketing Research, *published by the American Marketing Association Vol. viii (February 1971) pp. 67–70.*

2. To determine whether two sets of variables are statistically independent of one another in a linear sense, or conversely, to determine the magnitude of the relationships between the two sets.

3. To explain the nature of any relationships between the sets of variables, generally by measuring the relative contribution of each variable to the canonical relationships obtained.

The first objective is primarily one of prediction, and typically there are few problems of interpretation. Because interpretation may become much more difficult in the latter two instances, we have emphasized these. The next section treats determination of the magnitude of a canonical relationship, and the following section deals with interpreting the nature of a canonical relationship. Final sections are general comments and an example to illustrate the points discussed.

INTERPRETING CANONICAL ANALYSIS

The existence of relationships between two variable sets has traditionally been determined by testing the statistical significance of the canonical correlation coefficients. Several adequate tests exist, e.g., [2, 3, 15, 16, 27], so there is seldom any difficulty in interpreting the probability that the coefficients are significantly different from zero. Interpretation problems begin to mount when attempts are made to assess how strongly (beyond statistical significance) the two sets of variables are related in a practical sense. Since the canonical correlations are by definition maximal, canonical relationships between sets are invariably overstated. A common method of assessing relationship strength is to use the roots (squared canonical R's) as estimates of the shared variance between linear combinations of the variables in each set. Such an approach is directly analogous to squaring the multiple correlation coefficient in regression to measure the amount of variation in the dependent variable that is associated with variation in the independent variables. However, there is one immediate problem: how to deal with $c(c > 1)$ canonical correlations.

One possible solution of the problem involves extending a method for estimating the average canonical relationship between two data sets of nominal measurements [25]. This extension provides an index of the average relationship as the mean square of the canonical correlations (MSCC). If $R_i = i$th canonical correlation, and c correlations are obtained, then the unbiased estimate of the mean square of the canonical correlation can be computed by:

$$\text{MSCC} = \sum_{i=1}^{o} R_i^2/c.$$

Another measure of the overall relationship obtained by canonical analysis of two sets of variables is to establish the proportion of variance shared by linear composites of the two sets of variables; this proportion can be extracted by *any given number* of canonical relationships. This procedure is analogous to a measure in factor analysis of the percentage of total variance in a set of variables that is extracted by a given number of factors. Just as it might be desirable to know how much total variance can be extracted by factors with eigenvalues of 1.0 or more, one might also wish to know how much of the common variance in linear combinations of both sets is extracted by a subset of the canonical relationships (say those that are significant beyond some alpha-level, such as .05 or .01).

Each canonical root can be multiplied by the residual variances from which its variates were extracted, because: (1) the root reflects the variates' shared variance and (2) each successive relationship is orthogonal to the preceding one. Summing these products for all roots gives a measure of the association between sets of variables, taking into account the variance in linear composites of the sets extracted through each stage of the canonical analysis. The total variance extracted can be expressed as:

$$\text{TVE} = \sum_{j=1}^{n} \sum_{i=0}^{j-1} R_j^2\,(1 - R_i^2)$$

where n = number of canonical relationships extracted. Unfortunately, as appealing and simple as these kinds of associational measures seem, they may lead to misinterpretations of the relationships between the sets, since the roots represent variance shared by *linear composites* of the sets, not the unweighted variables themselves. In fact, computation of optimal weights in order to maximize correlations between linear composites of variables would imply that the original variables are not closely related. Thus "a relatively strong canonical correlation may obtain between two linear functions, even though these linear functions may not extract significant proportions of variance from their respective batteries" [26, p. 160].

Measuring Redundancy

To rectify such inherent overstatement in measures of canonical association, Stewart and Love [26], Ragland [21], and Miller and Farr [18, 19], apparently working independently, developed similar measures to assess the *average* relationship between two *sets* of variables. Because the work of Stewart and Love is probably the most accessible, their terminology is employed in the remainder of this section.

If A represents the variation in a set of variables measured over several objects, and B represents variation in another set of measures on the same objects, then $A \cap B$ may be taken as a measure of their shared variation. This intersection of the two sets is termed "redundancy" by Stewart and Love. Redundancy, expressed as a percentage of the total variation in each set, is rarely symmetrical in the sense that the percentages are the same for both variable sets, because the total variance and the number of variables in each set will differ as in the example below:

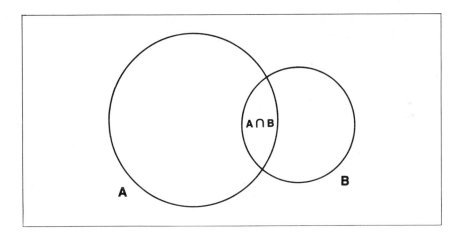

Since the smaller number of variables limits the number of canonical functions that may be extracted, the percentages of variance extracted from each set will also differ. Ordinarily, all the variance from the smaller set of variables will be extracted, but less from the larger set. In sum, since redundancy is not symmetric, an index that reflects this asymmetry is needed. To simplify, we may consider one set of variables as the predictor set (P) and the other as the criterion set (C), with the understanding that logically the relationship could be reversed and the same sense of prediction would hold.

Hotelling has shown that canonical analysis is the equivalent of performing independent principal components analyses on each of two sets of variables. Then the resulting component structures are rotated to develop weights for each variable that produce maximal correlations between components on each side [12]. Stewart and Love's redundancy technique uses this concept of a factor extracting variance from a set of variables; it also uses the notation of factor analysis. They define the sum of the squared correlations (loadings) of a canonical variate (or factor) with the individual variables within a set as the variance extracted by that variate (the communality of that factor). Dividing this sum by the number of variables in the set (M) gives the proportion of the set's variance extracted by that canonical variate. Thus:

VC_i = proportion of criterion set variance extracted by ith factor (or variate)

VP_i = proportion of predictor set variance extracted by ith factor

λ_i = squared ith canonical correlation coefficient, or the proportion of the variance of one of the ith pair of canonical variates predictable from the other member of the pair.

If VC_i is multiplied by λ_i, the product is the proportion of the variance of the C set variables explained by correlation between the ith pair of canonical variates. Calculating and summing this value for the M_c pairs of variates gives an index of the proportion of variance of C predictable from P, or the redundancy in C given P:

$$\overline{R^2}_{c/p} = \sum_{i=1}^{M_c} \lambda_i VC_i = \sum_{i=1}^{M_c} \lambda_i \left[\sum_{j=1}^{M} \frac{L_{ji}^2}{M} \right]$$

where L_{ji} is the correlation (loading) between the jth variable and the ith canonical variate (for the criterion side).

This index is termed $\overline{R}_{c/p}^2$ because it is the equivalent of computing the squared multiple correlation coefficient between the total P set and each variable in the C set, and then averaging these squared coefficients to arrive at an average R^2. This redundancy is the area of the intersection of the two sets of variables, expressed as a proportion of the C set. If all VP_i are computed for the P set, it is possible to derive an analogous index of redundancy for the P set given the C set, expressed as $\overline{R}_{p/c}^2$.

Therefore, redundancy provides a summary measure of the average ability of a set of P variables (taken as a set) to explain variation in C criterion variables (taken one at a time). Rather than relying on an "inflated" canonical root, which reflects an optimum weighting of both sides of the relationship, we predict actual scores for criterion variables and measure the accuracy of this prediction.

In sum, it seems reasonable to use canonical correlation coefficients to test for the existence of overall relationships between sets of variables, but for a measure of the magnitude of the relationships, redundancy may be more appropriate.

Explaining the Nature of Canonical Relationships

The next step is to explain the relative importance of each of the variables in determining the relationships. Typically the nature of canonical relationships has been inferred by noting the sign and magnitude of the standardized canonical weight assigned each variable in computing canonical functions. While weights are useful for predictive accuracy, interpreting the relative importance or contribution of a variable by its canonical weight has all the difficulties associated with the interpretation of beta weights in regression, where independent variables are correlated. Unless the variables within each set have been previously reduced to linearly independent variates (or possibly, to relatively independent variables) by factor analysis, multicollinearity may obscure the relationships or render results unstable because of sampling error. On the one hand, a small weight may mean that its corresponding variable is irrelevant in determining a relationship. On the other hand, it may mean that the variable has been partialed out of the relationship because it possesses a high correlation with another variable that correlates slightly higher with the respective canonical variate of the other set of variables. Although not necessary for prediction purposes, such a variable should definitely be acknowledged as an important part of the canonical relationship.

For better interpretation, within-set variable-variate correlations may be computed for each (independent) canonical relationship. This within-set variance should be treated as analogous to a factor analysis problem in which both factor and variable communalities are computed [17, 21, 27]. However, such computations only provide information about the relative contributions of variables to *each* independent canonical relationship. As yet there is no direct estimate of the relative contribution of a variable to the *overall* canonical relationship.

Another analytical tactic is to express the squared factor loadings as percentages of the variance extracted by their respective canonical variates. By such a manipulation, variables may be identified as important to the structure of a particular canonical factor even though the loading is small in an absolute sense [21].

Accordingly, when attempting to explicate canonical relationships, each one should be explored independently and examined on both sides of the relationship for clusters of variables that are highly loaded (say, relative to the appropriate sum of squared loadings) on their factors.

An additional question arises concerning which canonical relationships should be explicated by interpreting the variable loadings, and which should be left "uninterpreted." The most common practice is to analyze only those functions whose canonical correlation coefficients are statistically significant at some preselected level, typically .05. The variables in each set which contribute heavily to shared variable-variate variance for these functions are then said to be related to each other. However, some researchers may prefer to view the proportion of total redundancy accounted for by a canonical factor as a better measure of whether or not its variables should be interpreted. In this case, they interpret functions in each set of variables that contribute substantially to redundancy with the other set, and "discard" relationships which do not. This procedure would be analogous to truncating a stepwise regression at a point when additional variables do not substantially increase R^2. Patterns of variables in relationships that are nonsignificant or contribute little to explained redundancy are not as relevant to the relationships between sets as are those patterns involved in more significant or higher redundancy-explaining relationships.

TABLE 1
COMPONENTS OF REDUNDANCY MEASURE

Relationship	Canonical R	R² (or λ)	Variance extracted, VP or VC	Redundancy (R²), λ :VP, or λ :VC	Proportion of total redundancy
Predictor set					
1	.6180[a]	.3819	.2999	.1146	.6753
2	.4104[a]	.1684	.2481	.0418	.2463
3	.2510[a]	.0630	.2082	.0131	.0772
4	.0458	.0021	.0856	.0002	.0012
				.1697	
Criterion set					
1	.6180[a]	.3819	.2994	.1143	.6978
2	.4104[a]	.1684	.2174	.0366	.2234
3	.2510[a]	.0630	.1451	.0123	.0751
4	.0458	.0021	.2881	.0006	.0036
				.1638	

[a] $p < .02$.

Further Comments

A high bivariate correlation between a variable in the C set and one in the P set leads to a spuriously high canonical correlation. The two variables correlating significantly will be assigned high canonical weights, and remaining variables support this relationship [8, p. 350]. Hence it is still useful to investigate all interset bivariate correlations for clearest interpretation of the results.

Also, since canonical correlations tend to be unstable, it is wise to estimate correlations and weights with one sample and cross-validate them with a second sample. Such cross-validation serves as an empirical test of statistical inference; by repeated sample-splitting and replication much more information about the data may be gained. However, when the sample is small (as a multiple of the total number of variables), cross-validation may use more information than it contributes.

Note, however, that the indices of redundancy are not particularly sensitive to the occasional overweighting of variables whose errors of measurement correlate by chance with the "dependent" variates. Averages are taken across all variables, including those not highly weighted in the composite functions. Thus, while the λ's reflect an upward bias without cross-validation, the $R_{c/p}^2$'s are still accurate estimates of the proportion of variance in a group of variables predictable from a second set of variables.

EXAMPLE

A heterogeneous sample of 196 male and female adults provided demographic data and data on consumption of cold cereal and beer and frequency of eating out and attending movies. Canonical analysis was performed on these data to determine whether any relationships existed between these data sets and, if so, the nature of these relationships. Demographic variables were treated as predictor variables; consumption variables were considered criterion variables.

Of four canonical correlations, three were significant beyond the .02 level (Table 1). It is all too easy to infer that demographics account for a great proportion of the underlying variance in consumption habits. For instance, the first canonical relationship alone appears to have accounted for over 38% of the common variance. However, this figure represents only the shared variation between *linear composites* of the variables.

Redundancy in the criterion set (given the predictor set) was computed to present a more realistic view of the shared variance. Note that the total redundancy was 16.38% of the variance in the criterion set, with the three significant relationships accounting for all but .06% of this amount. While the redundancy still indicates a substantial amount of shared variance, the adjustment markedly lowers the shared variance estimate originally obtained by correlating the optimal linear combinations.[2]

Table 2 presents three measures of the contribution of each variable to the canonical relationships. The first measure is the standardized form of the canonical weight of the variables used in determining each subject's canonical scores. The second measure is the correlation between individual variables and the respective canonical variates; these corresponded to loadings (L) in factor analysis. The final measure is the variable-variate correlation squared and expressed as a percentage of the sum of squared correlations for each variate. These three measures usually provide sufficient information to interpret canonical functions.

Comparing the relative importance of the variables as indicated by the three measures reveals the difficulties which might occur in designating variables as "highly" related to each canonical relationship. In general, although the rank order of importance was not much affected, some differences were indicated. For example, using weights, the rank order of demographics contributing to the first relationship was age, sex, education, marital status, and income. However, using loadings, the order was age, education, marital status, sex, and income. Thus although all these variables (e.g., with loadings below .30) would not be investigated, the differences in ranking are important. Similar changes appeared for the consumption variables.

Use of the squared loadings also is a useful device in interpretation. If all squared loadings for a variable summed to 1.0, this method would have no advantage over using absolute loadings, since results would be identical throughout the matrix. However, since factor communalities may differ for different relationships (and for different sides of the same relationship), a variable might be seen as critical to one relationship when the same loading would not qualify it for inclusion in some other relationship. For instance, in the case of the third relationship, loadings for education and income were higher than that for restaurants. However, variance in restaurant usage was

---

[2] Total variance extracted = .3819 + .1684 (1 − .3819) + . . . = .5194. Also, the modified Srikantan index of the mean square of the canonical coefficient yields: $\Sigma R_i^2/C = .1538$.

TABLE 2
RELATIONSHIPS BETWEEN VARIABLES AND CANONICAL FUNCTIONS

Variables	Canonical relationships								
	1			2			3		
	Weight	Loading	Percentage L^2	Weight	Loading	Percentage L^2	Weight	Loading	Percentage L^2
Predictor set (demographics)									
Sex[a]	−.439	−.451	13.6	−.781	−.763	46.9	.616	.445	19.0
Marital status[b]	−.105	−.474	15.0	.067	−.049	.2	−.116	.070	.5
Age	−.825	−.855	48.8	.310	.434	15.2	.037	−.010	.0
Education	.338	.576	22.1	−.013	.121	1.2	.530	.637	39.0
Income	−.033	−.089	.5	.539	.674	36.6	.570	.658	41.6
			100.0			100.1			100.1
Criterion set (product use)									
Cereal	.102	.070	.4	.340	.385	17.1	.513	.379	18.4
Beer	.520	.614	31.4	.590	.621	44.3	−.553	−.488	30.5
Movies	.843	.864	62.3	−.583	−.468	25.2	.155	.187	4.5
Restaurants	.095	.265	5.9	.443	.342	13.4	.639	.603	46.6
			100.0			100.0			100.0

[a] Males were coded "1" and females "2."
[b] Single was coded "1" and married "2."

more closely associated with criterion variance extracted by the third variate than the corresponding figures indicated for income and education, relative to variance extracted by the third predictor set variate. Here the differences were minor, but in other instances they may be considerable. Thus rank order of variable importance for any side of a canonical relationship does not depend upon whether loadings or percentage of the sum of squared loadings is used (a monotonic transformation). However, the order may change when both sides of a relationship or several relationships are considered. The measure of the percentage of variate communality associated with each variable is probably more sensitive to variations in contributions of variables than is the simple choice of constant loading cutoff. Both measures should probably be used simultaneously, so that the signs of the loadings yield directional information about variables' contributions to the canonical relationships.[3]

CONCLUSIONS

When correctly applied, canonical analysis is a highly useful and powerful technique for exploring the nature and strengths of relationships between sets of variables. Its use in marketing should increase as researchers learn more about its potentialities and how to interpret its results. Interpretation of canonical analysis will be enhanced by considering the following conclusions.

First, the squared canonical correlation coefficients (roots) reflect shared variation between linear composites of the original variables, *not* the variables themselves. For a measure of the shared variation between the sets of variables, compute the redundancy between the sets, redundancy being the equivalent of the average multiple correlation coefficient squared between one set and variables in the other set.

Second, depending on the purposes of the study, interpretation of the relative contribution of the variables to each canonical relationship may require consideration of their correlations with the canonical variates, rather than assessing importance from vectors of standardized canonical weights. Weights appear more suitable for

---

[3] Signs may be translated for ease of expression. However, while a relatively low value for age (for example) would contribute to a high value of the first demographic variate, one cannot conclude that youths drink a lot of beer. One must say that for this sample relatively low age was associated with relatively high consumption of beer. The same caution is needed in interpreting loadings, weights, and percentage of $\Sigma\ L^2$.

prediction, while correlations may better explain underlying (although interrelated) constructs.

Third, each pair of canonical variates is independent from previous (and succeeding) pairs. Hence a different relationship arises from each one. The presence of some of the same key variables in two or more pairs will still yield a different "whole" when the other variables that are highly correlated with the same canonical variate are also considered.

Finally, determining cut-off points for including variables requires subjective judgment and depends on the purposes of the analysis. Care must be taken not to "twist the data."

With these guides in mind, canonical analysis can be creatively applied to many marketing problems in both a predictive and an explanatory sense. The suggestions of Green and Tull [7] and Horton, Russell, and Moore [10] to apply canonical analysis in conjunction with other multivariate techniques is an excellent one. In this way one might factor analyze variables to remove multicollinearity within variable sets prior to canonical analysis. Moreover, other multivariate techniques could be used to validate tentative conclusions reached through an exploratory canonical analysis, to determine if other ways of analyzing the data produce comparable conclusions.

REFERENCES

1. Anderson, T. W. *Introduction to Multivariate Statistical Analysis*. New York: John Wiley & Sons, 1958.
2. Bartlett, M. S. "The Statistical Significance of Canonical Correlations," *Biometrika*, 32 (January 1941), 29–38.
3. ———. "Multivariate Analysis," *Journal of the Royal Statistical Society, Supplement*, 9 (1947), 176–90.
4. Cooley, William A. and Paul R. Lohnes. *Multivariate Procedures for the Behavioral Sciences*. New York: John Wiley & Sons, 1962.
5. De Groot, M. H. and E. C. C. Li. "Correlations Between Similar Sets of Measurements," *Biometrics*, 22 (December 1966), 781–90.
6. Gower, J. C. "A Q-technique for the Calculations of Canonical Variates," *Biometrika*, 53 (December 1966), 588–90.
7. Green, Paul E. and Donald S. Tull. *Research for Marketing Decisions*, second edition. Englewood Cliffs, N.J.: Prentice-Hall, 1970, Chapter 11.
8. Gullikson, H. *Theory of Mental Tests*. New York: John Wiley & Sons, 1950.
9. Horst, Paul. "Relations Among *M* Sets of Measures," *Psychometrika*, 26 (June 1961), 129–49.
10. Horton, I. F., J. S. Russel, and A. W. Moore. "'Multivariate Covariance and Canonical Analysis: A Method of Selecting the Most Effective

Discriminators in a Multivariate Situation," *Biometrics,* 24 (December 1968), 845–58.

11. Hotelling, Harold. "The Most Predictable Criterion," *Journal of Educational Psychology,* 26 (February 1935), 139–42.

12. ———. "Relations Between Two Sets of Variates," *Biometrika,* 28 (December 1936), 321–77.

13. Kendall, Maurice G. *A Course in Multivariate Analysis.* New York: Hafner, 1957.

14. Koons, Paul B., Jr. "Canonical Analysis," in Harold Borko, ed., *Computer Applications in the Behavioral Sciences.* Englewood Cliffs, N.J.: Prentice-Hall, 1962, 266–79.

15. Lawley, D. N. "Tests of Significance in Canonical Analysis," *Biometrika,* 46 (June 1959), 59–66.

16. Mariott, F. H. C. "Tests of Significance in Canonical Analysis," *Biometrika,* 39 (May 1952), 58–64.

17. Meredith, William. "Canonical Correlations with Fallible Data," *Psychometrika,* 29 (March 1964), 55–65.

18. Miller, John K. "The Development and Application of Bimultivariate Correlation: A Measure of Statistical Association Between Multivariate Measurement Sets," unpublished doctoral dissertation, State University of New York at Buffalo, 1969.

19. ——— and S. David Farr. "Bimultivariate Redundancy: A Comprehensive Measure of Interbattery Relationship," *Multivariate Behavioral Research,* 6 (July 1971), 313–24.

20. Morrison, Donald E. *Multivariate Statistical Methods.* New York: McGraw-Hill, 1967.

21. Ragland, Robert E. "On Some Relations Between Canonical Correlation, Multiple Regression, and Factor Analysis," unpublished doctoral dissertation, University of Oklahoma, 1967.

22. Roskam, E. "A Program for Computing Canonical Correlations on IBM 1620," *Educational and Psychological Measurement,* 26 (Spring 1966), 193–8.

23. Roy, S. N. *Some Aspects of Multivariate Statistics.* New York: John Wiley & Sons, 1957.

24. Rozeboom, W. W. "Linear Correlations Between Sets of Variables," *Psychometrika,* 30 (March 1965), 57–71.

25. Srikantan, K. S. "Canonical Association Between Nominal Measurements," *Journal of the American Statistical Association,* 65 (March 1970), 284–92.

26. Stewart, Douglas and William Love. "A General Canonical Correlation Index," *Psychological Bulletin,* 70 (September 1968), 160–3.

27. Veldman, Donald J. *Fortran Programming for the Behavioral Sciences.* New York: Holt, Rinehart and Winston, 1967.

6

Factor Analysis

CHAPTER REVIEW

The multivariate statistical technique of factor analysis has found increased use during the past decade in the various fields of business related research, especially in marketing and personnel management. The purpose of this chapter is to describe factor analysis, a technique particularly suitable for analyzing the complex, multi-dimensional problems encountered by researchers and businesspeople. This chapter defines and explains in broad conceptual terms the fundamental aspects of factor analytic techniques. Basic guidelines for presenting and interpreting the results of using these techniques also are included to further clarify the methodological concepts. Before proceeding further, it will be helpful for you to review the Definitions of Key Terms.

Factor analysis can be utilized to examine the underlying patterns or relationships for a large number of variables, and determine if the information can be condensed or summarized in a smaller set of factors or components. An understanding of the most important factor analysis concepts should enable you to:

- [ ] Differentiate factor analytic techniques from other multivariate techniques.
- [ ] State the major purposes of factor analytic techniques.
- [ ] Identify the difference between component analysis and common factor analysis models.
- [ ] Tell when component analysis and common factor analysis should be utilized.
- [ ] Identify the difference between "R" factor analysis and "Q" factor analysis.
- [ ] Explain the concept of rotation of factors.
- [ ] Tell how to determine the number of factors to extract.
- [ ] Explain the purpose of factor scores and how they can be used.
- [ ] Explain how to select surrogate variables for subsequent analysis.
- [ ] State the major limitations of factor analytic techniques.

DEFINITIONS OF KEY TERMS

CORRELATION MATRIX. A table showing the intercorrelations among all variables.

FACTOR. A linear combination of the original variables. Factors also represent the underlying dimensions (constructs) that summarize or account for the original set of observed variables.

FACTOR MATRIX. A table displaying the factor loadings of all variables on each of the factors.

FACTOR LOADINGS. The correlation between the original variables and the factors, and the key to understanding the nature of a particular factor. *Squared factor loadings* indicate what percent of the variance in an original variable is explained by a factor.

COMMUNALITY. The amount of variance an original variable shares with all other variables included in the analysis.

FACTOR ROTATION. The process of manipulating or adjusting the factor axes in a clockwise direction to achieve a simpler and theoretically more meaningful factor solution.

COMPONENT ANALYSIS. A factor model in which the factors are based upon the total variance. With component analysis, unities are inserted in the diagonal of the correlation matrix.

COMMON FACTOR ANALYSIS. A factor model in which the factors are based upon a reduced correlation matrix. That is, communalities are inserted in the diagonal of the correlation matrix, and the extracted factors are based only on the common variance, with specific and error variance excluded.

ORTHOGONAL. Refers to the mathematical independence of factor axes to each other (i.e., at right angles or 90 degrees).

ORTHOGONAL FACTOR SOLUTIONS. A factor solution in which the factors are extracted so that the factor axes are maintained at 90 degrees. Thus, each factor is independent or orthogonal from all other factors. The correlation between factors is arbitrarily determined to be zero.

OBLIQUE FACTOR SOLUTIONS. A factor solution computed so that the extracted factors are correlated. Rather than arbitrarily constraining the factor solution so the factors are orthogonally independent to each other, the analysis is conducted to express the actual relationship between the factors which may or may not be orthogonal.

EIGENVALUE. The column sum of squares for a factor; also referred to as the latent root. It represents the amount of variance accounted for by a factor.

TRACE. The total amount of variance the factor solution is based upon. With component analysis, the trace is equal to the number of variables based on the assumption that the variance in each variable is equal to one. With common factor analysis, the trace is equal to the sum of the communalities on the diagonal of the reduced correlation matrix (also equal to the amount of common variance for the variables being analyzed).

6

WHAT IS FACTOR ANALYSIS?

Factor analysis is a generic name given to a class of multivariate statistical methods whose primary purpose is data reduction and summarization. Broadly speaking, it addresses itself to the problem of analyzing the interrelationships among a large number of variables (e.g., test scores, test items, questionnaire responses), and then explaining these variables in terms of their common, underlying dimensions (factors). For example, a hypothetical survey questionnaire may consist of 100 questions; but since not all of the questions are identical, they do not all measure the basic underlying dimensions to the same extent. By using factor analysis, the analyst can identify the separate dimensions being measured by the survey, and determine a factor loading for each variable (test item) on each factor.

Factor analysis (unlike multiple regression, discriminant analysis or canonical correlation, in which one or more variable is explicitly considered the criterion or dependent variable, and all others the predictor or independent variables), is an interdependence technique in which all variables are simultaneously considered. In a sense, each of the observed (original) variables is considered as a dependent variable that is a function of some underlying, latent and hypothetical set of factors (dimensions). Conversely, one can look at each factor as a dependent variable that is a function of the originally observed variables.

PURPOSES OF FACTOR ANALYSIS

The general purpose of factor analytic techniques is to find a way of condensing (summarizing) the information contained in a number of original variables into a smaller set of new composite dimensions (factors) with a minimum loss of information. That is, to search for and define the fundamental constructs or dimensions assumed to underlie the original variables. More specifically, four functions factor analysis techniques can perform are [6]:

(1) Identify a set of dimensions that are latent (not easily observed) in a large set of variables; also referred to as "R" factor analysis.

(2) Devise a method of combining or condensing large numbers of people into distinctly different groups within a larger population; also referred to as "Q" factor analysis.

(3) Identify appropriate variables for subsequent regression, correlation or discriminant analysis from a much larger set of variables.

(4) Create an entirely new set of a smaller number of variables to partially or completely replace the original set of variables for inclusion in subsequent regression, correlation or discriminant analysis.

Approaches (1) and (2) take the identification of the underlying dimensions or factors as ends in themselves; the estimates of the factor loadings are all that is required for the analysis. Method (3) also relies on the factor loadings; but uses them as the basis for identifying variables for subsequent analysis with other techniques. Method (4) requires that estimates of the factors themselves (factor scores) be obtained; then the factor scores are used as independent variables in a regression, discriminant, or correlation analysis.

Factor Analysis Decision Diagram

Figure 6.1 shows the general steps followed in any application of factor analysis techniques. The starting point in factor analysis, as with other statistical techniques, is the research problem. If the objective of the research is data reduction and summarization, then factor analysis is the appropriate technique to use. Questions which the analyst needs to answer at this point are: what variables should be included, how many variables should be included, how are the variables measured, and is the sample size sufficiently large enough? Regarding the question of variables, any variables relevant to the research problem can be included as long as they are appropriately measured. Raw data variables for factor analysis are generally assumed to be of metric measurement. In some cases, dummy variables (coded zero-one) although considered non-metric can be used. Regarding the sample size question, the researcher generally would not factor analyze a sample of less than 50 observations, and preferably the sample size should be 100 or larger. As a general rule there should be four or five times as many observations as there are variables to be analyzed. This ratio is somewhat conservative, and in many instances the researcher is forced to factor analyze a set of variables when only a ratio of twice the observations to the number of variables is available. When dealing with smaller sample sizes and a lower ratio, the analyst should cautiously interpret any findings.

One of the first decisions in the application of factor analysis involves the calculation of the correlation matrix. Based upon the research problem, the analyst must define the relevant universe for

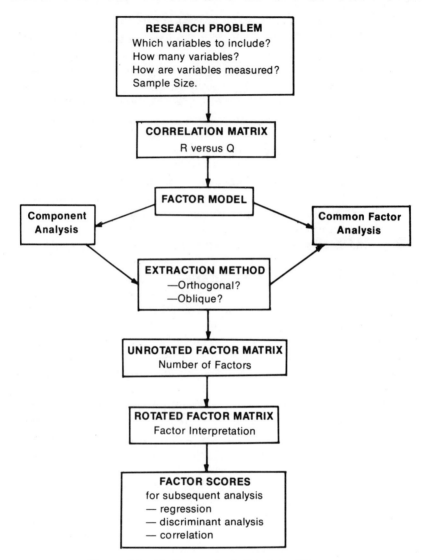

Figure 6.1—Factor Analysis Decision Diagram

analysis. The alternative would be to either examine the correlations between the variables or the correlations between the respondents. For example, suppose you have data on 100 respondents in terms of ten characteristics. It is possible to calculate the correlations between each of the ten characteristics or between each of the individuals. If the objective of the research is to summarize the characteristics, then the factor analysis would be applied to a correlation matrix of the variables. This is the most common type of factor analysis, and is referred to as "R" factor analysis. It corresponds to the first purpose of factor analysis outlined previously. Factor analysis also may be applied to a correlation matrix of the individual respondents. This type of analysis, called "Q" factor analysis, was previously identified as a second possible purpose of factor analysis. A "Q" factor analysis approach is not utilized very frequently. Instead, most analysts will utilize some type of cluster analysis or hierarchical grouping technique to group individual respondents.

Numerous variations of the general factor model are available. The two most frequently employed factor analytic approaches are component analysis [1] and common factor analysis. Selection of the factor model depends upon the analyst's objective. The component model is used when the objective is to summarize most of the original information (variance) in a minimum number of factors for prediction purposes. In contrast, common factor analysis is used primarily to identify underlying factors or dimensions not easily recognized. Both of these factor models will be discussed in more detail in the following sections.

In addition to selecting the factor model, the analyst must specify how the factors are to be extracted. Two options are available, either orthogonal factors or oblique factors. In an orthogonal solution, the factors are extracted in such a way that the factor axes are maintained at 90 degrees, meaning that each factor is independent from all other factors. Therefore the correlation between factors is arbitrarily determined to be zero. An oblique factor solution is more complex than is an orthogonal one. In fact, an entirely satisfactory analytical procedure has not been devised for oblique solutions. They are still the subject of considerable experimentation and controversy [3]. As the term oblique implies, the factor solution is computed so that the extracted factors are correlated. Oblique solutions assume the original variables or characteristics are correlated to some extent, therefore the underlying factors must be similarly correlated. To summarize, orthogonal factor solutions are mathematically simpler to handle, while oblique factor solutions are more flexible and more realistic, because the theoretically important underlying dimensions are not assumed to be unrelated to each other.

The choice of an orthogonal or oblique rotation should be made on the basis of the particular needs of a given research problem. If

---

[1] Many texts refer to this approach as principal components. For our purposes component analysis is the same as principle components analysis.

the goal of the research is to reduce the number of original variables regardless of how meaningful the resulting factors may be, then the appropriate solution would be an orthogonal one. Also, if the researcher wants to reduce the larger number of variables into a smaller set of uncorrelated variables for subsequent use in a regression or other prediction technique, then an orthogonal solution is the best. However, if the ultimate goal of the factor analysis is to obtain several theoretically meaningful factors or constructs, then an oblique solution is appropriate. This is because realistically, very few variables are uncorrelated as in an orthogonal solution.

When a decision has been made on the correlation matrix, the factor model, and the extraction method, the analyst is then ready to extract the initial unrotated factors. By examining the unrotated factor matrix, the analyst can explore the data reduction possibilities for a set of variables and obtain a preliminary estimate of the number of factors to extract. Final determination of the number of factors must wait, however, until the factor matrix is rotated and the factors are interpreted.

Depending upon the objective for applying factor analysis techniques, the research may stop with factor interpretation or proceed on to the computation of factor scores and subsequent analysis with other statistical techniques. If the objective is simply to identify logical combinations of variables or respondents (purposes 1 and 2), then the analyst will stop with the factor interpretation. If the objective is to identify appropriate variables for subsequent application to other statistical techniques (purpose 3), then the research would examine the factor matrix and select the variable with the highest factor loading as a surrogate representative for a particular factor dimension. If the researcher's objective is to create an entirely new set of a smaller number of variables to replace the original set of variables for inclusion in a subsequent type of statistical analysis (purpose 4), then composite factor scores would be computed to represent each of the factors. The factor scores would then be used as the raw data to represent the independent variables in a regression, discriminant or correlation analysis.

Approaches for Deriving the Correlation Matrix

As noted previously, one of the first decisions in the application of factor analysis focuses on the approach to calculating the correlation matrix. The analyst could derive the correlation matrix based on the computation of correlations between the variables. This would be an "R" type factor analysis, and the result would be a factor pattern demonstrating the underlying relationships of the variables. The analyst could also elect to derive the correlation matrix based on the correlations between the individual respondents. This is referred to as "Q" type factor analysis, and the results would be a factor matrix which would identify similar individuals. For example, if the individual respondents are identified by number, then the resulting factor pattern might tell you that individuals 1, 5, 7 and 10 are similar. These respondents would be grouped together because they exhibited a high loading on the same factor. Similarly, respondents 2, 3, 4 and 8 would perhaps load together on another factor. We would label these individuals as being similar. From the results of a "Q" factor analysis we could identify groups or clusters of individuals demonstrating a similar response pattern on the variables included in the analysis.

A logical question at this point would be: "How does "Q" type factor analysis differ from cluster analysis?" The answer is that both approaches compare a series of responses to a number of variables and place the respondents into several groups. The difference would be that the resulting groups for a "Q" type factor analysis would be based on the intercorrelations between the means and standard deviations of the respondents. The cluster analysis approach would devise its groupings based on the absolute distances between the respondents' scores on the variables being analyzed [10]. To illustrate this difference consider Table 6.1. This table contains the scores of four respondents over three different variables. A "Q" type factor analysis of these four respondents would yield two groups with similar variance structures. The two groups would consist of respondents A and C versus B and D. In contrast, the clustering approach would be sensitive to the absolute distances among the respondents' scores and would lead to a grouping of the closest pairs. Thus, with a cluster

TABLE 6.1
COMPARISONS OF SCORE PROFILES FOR Q-TYPE FACTOR
ANALYSIS AND HIERARCHICAL CLUSTER ANALYSIS

Respondent	(1)	(2)	(3)
A	7	6	7
B	6	7	6
C	4	3	4
D	3	4	3

analysis approach respondents A and B would be placed in one group and C and D would be in the other group [10].

Common Factor Analysis and Component Analysis

There are two basic models the analyst can utilize to obtain factor solutions. They are known as common factor analysis and component analysis. To select the appropriate model, the analyst must understand something about the types of variance. For the purposes of factor analysis, total variance consists of three kinds: (1) common, (2) specific, and (3) error. These types of variance and their relationship to the factor model selection process are illustrated in Figure 6.2. **Common variance** is defined as that variance in a variable which is shared with all other variables in the analysis. **Specific variance** is that variance associated with only a specific variable; and **error variance** is that due to unreliability in the data gathering process or a random component in the measured phenomenon. When using component analysis, the total variance is considered and hybrid factors are derived which contain small proportions of unique and in some instances error variance, but not enough in the first few factors to distort the overall factor structure. Specifically, with component analysis unities are inserted in the diagonal of the correlation matrix. Conversely, with common factor analysis communalities are inserted in the diagonal, and the factors are derived based only on the common variance. From a variance point of view, there is a big difference between inserting unity in the diagonal or using communality estimates. With unity in the diagonal, the full variance is brought into the factor matrix as shown in Figure 6.2. Common factor analysis substitutes communality estimates in the diagonal, and the resulting factor solution is based only on common variance.

The common factor and component analysis models are both widely utilized. The selection of one model over the other is based upon two criteria: (1) the objective of the researcher conducting the factor analysis, and (2) the amount of prior knowledge about the variance in the variables. When the analyst is primarily concerned about: prediction, determining the minimum number of factors needed to account for the maximum portion of the variance represented in the original set of variables, and has prior knowledge suggesting that unique and error variance represent a relatively small proportion of the total variance, then the appropriate model to select is the component analysis model. In contrast, when the primary objective is to identify the latent dimensions or constructs represented in the original variables, and the researcher has little knowledge about the

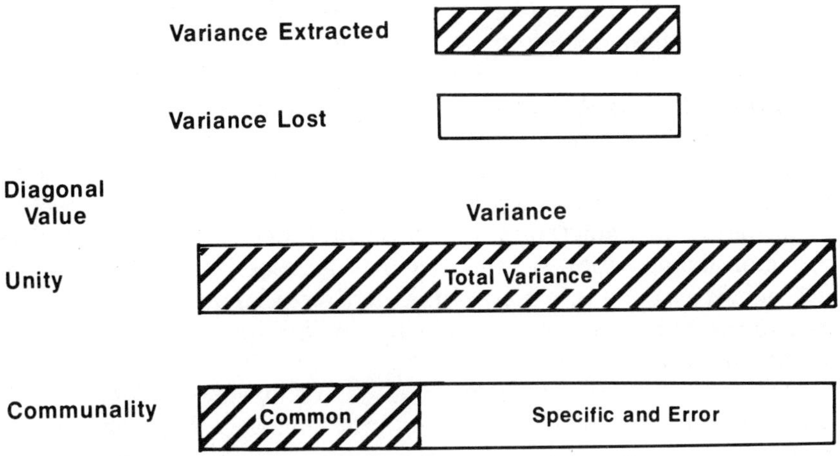

Variance Extracted

Variance Lost

Diagonal
Value

Variance

Unity

Total Variance

Communality

Common

Specific and Error

Figure 6.2—Types Of Variance Carried Into Factor Matrix

amount of unique or error variance and therefore wishes to eliminate this variance, the appropriate model to select is the common factor model. It is important to note at this point that either model can be selected and applied with relative ease. This is made possible by computers which derive good approximations of communalities through repeated calculations.

The Rotation of Factors

An important concept in factor analysis is the rotation of factors. The term "rotation" in factor analysis means exactly what it implies. Specifically, the reference axes of the factors are turned about the origin until some other position has been reached [3]. The simplest case is an orthogonal rotation in which the axes are maintained at 90 degrees. It is possible to rotate the axes and not retain the 90 degree angle between the reference axes. This rotational procedure is referred to as an oblique rotation. Orthogonal and oblique factor rotations are demonstrated using a graphical approach in Figures 6.3 and 6.4.

As was pointed out in the factor analysis decision diagram, two stages are involved in the derivation of a final factor solution. First, the initial unrotated factor matrix is computed to assist in obtaining a preliminary indication of the number of factors to extract. In computing the unrotated factor matrix, the analyst is simply interested in the best linear combination of variables—best in the sense that the particular combination of original variables would account for more of the variance in the data as a whole than any other linear combination of variables. Therefore, the first factor may be viewed as the single best summary of linear relationships exhibited in the data. The second factor is defined as the second best linear combination of the variables subject to the constraint that it is orthogonal to the first factor. To be orthogonal to the first factor, the second one must be derived from the proportion of the variance remaining after the first factor has been extracted. Thus, the second factor may be defined as the linear combination of variables that accounts for the most residual variance after the effect of the first factor is removed from the data. Subsequent factors are defined similarly until all the variance in the data is exhausted.

Unrotated factor solutions achieve the objective of data reduction, but the analyst must ask if the unrotated factor solution (while fulfilling desirable mathematical requirements), will provide information which offers the most adequate interpretation of the variables under examination. In most instances the answer to this question will be no. Therefore, the basic reason for employing a rotational method is to achieve simpler and theoretically more meaningful factor solutions. Rotation of the factors in most cases improves the interpretation by reducing some of the ambiguities which often accompany initial unrotated factor solutions [8].

The unrotated factor solution may or may not provide a meaningful patterning of variables. If the unrotated factors are expected to be meaningful, the user may specify that no rotation be performed. Generally, rotation will be desirable because it simplifies the factor structure and usually it is difficult to determine whether unrotated factors will be meaningful or not.

As indicated earlier, unrotated factor solutions extract factors in the order of their importance. The first factor tends to be a general factor with almost every variable loading significantly, and it accounts for the largest amount of variance. The second and subsequent factors will then be based upon the residual amount of variance. Each will account for successively smaller portions of variance. The ultimate effect of rotating the factor matrix is to redistribute the variance from earlier factors to later factors to achieve a simpler, theoretically more meaningful factor pattern.

To illustrate the concept of factor rotation, examine Figure 6.3 in which five variables are depicted in a two-dimensional factor diagram. The vertical axis represents the unrotated factor II and the horizontal axis represents unrotated factor I. The axes are labeled with a 0 at the origin and extending outward up to a +1.0 or a −1.0.

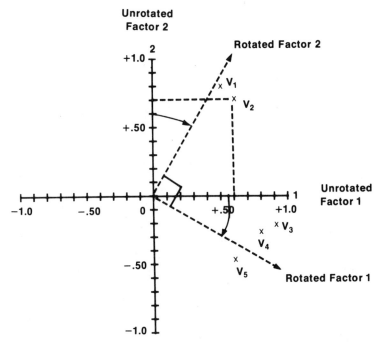

Figure 6.3—Orthogonal Factor Rotation

These numbers represent the factor loadings. The variables are labeled V_1, V_2, V_3, V_4, and V_5. The factor loading for variable 2 on the
unrotated factor II would be determined by drawing a dashed line
horizontally to the vertical axis for factor II. Similarly, a vertical one
would be drawn from variable 2 to the horizontal axis of the unrotated factor I in order to determine the loading of variable 2 on
factor I. A similar procedure would be followed for the remaining
variables until all the loadings are determined for all factor variables.
The factor loadings for the unrotated and rotated solutions are displayed in Table 6.2 for comparison purposes. On the unrotated first
factor all the variables load fairly high. On the unrotated second
factor, variables one and two are very high in the positive direction.
Variable five is moderately high in the negative direction, while variables three and four have considerably lower loadings in the negative
direction.

From visual inspection of Figure 6.3 it is obvious there are two
clusters of variables. Variables 1 and 2 go together, as well as variables
3, 4 and 5. However, such patterning of variables is not so obvious

TABLE 6.2
COMPARISON BETWEEN ROTATED AND UNROTATED
FACTOR LOADINGS

Variable	Unrotated Factor Loadings		Rotated Factor Loadings	
	I	II	I	II
V_1	.50	.80	.03	.94
V_2	.60	.70	.16	.90
V_3	.90	−.25	.95	.24
V_4	.80	−.30	.84	.15
V_5	.60	−.50	.76	−.13

from the unrotated factor loadings. By rotating the original axes clockwise as indicated in Figure 6.3, we obtain a completely different factor loading pattern. Note that in rotating the factors the axes are maintained at 90 degrees. This signifies the factors are mathematically independent and that the rotation has been orthogonal. After rotating the factor axes, variables 3, 4, and 5 load very high on factor one, and variables 1 and 2 load very high on factor two. Thus, the clustering or patterning of these variables into two groups is more obvious after the rotation than before, even though the relative position or configuration of the variables remains unchanged.

The same general principles pertain to oblique rotations as apply to orthogonal rotations. The oblique rotation method is more flexible because the factor axes need not be orthogonal. It also is more realistic because the theoretically important underlying dimensions are not assumed to be uncorrelated to each other. In Figure 6.4 the two rotational methods are compared. Note the oblique factor rotation represents the clustering of variables more accurately. This is because each rotated factor axis is now closer to the respective group of variables. Also, the oblique solution provides us with information about the extent to which the factors are actually correlated with each other.

Most factor analysts agree that many direct unrotated solutions are not sufficient. That is, in most cases rotation will improve the interpretation by reducing some of the ambiguities which often accompany the preliminary analysis. The major option available to the analyst in rotation is to choose an orthogonal method or an oblique method. The ultimate goal of any rotation is to obtain some theoretically meaningful factors, and if possible the simplest factor struc-

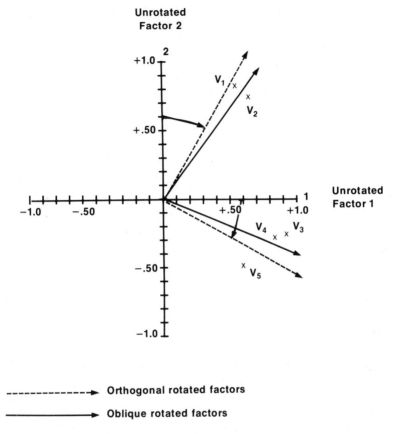

Figure 6.4—Oblique Factor Rotation

ture. Orthogonal rotational approaches are more widely used, because all computer packages performing factor analysis contain orthogonal rotation options. Only a few computer packages contain the oblique rotational option. Orthogonal rotations are also utilized more frequently, because the analytical procedures for performing oblique rotations are not as well developed and are still subject to considerable controversy. When the objective is to utilize the factor results in some kind of subsequent statistical analysis, then the analyst would always select an orthogonal rotation procedure. This is because the factors are orthogonal and therefore eliminate collinearity. However, if the analyst is simply interested in obtaining theoretically meaningful constructs or dimensions, then the oblique factor rotation is more desirable because it is theoretically and empirically more realistic.

One final topic needs to be discussed regarding the rotation of factors. Several different approaches are available for performing either orthogonal or oblique rotations. Only a limited number of oblique rotational procedures are available and the analyst will probably be forced to accept the one which is accessible. Since none of the oblique solutions have been demonstrated to be analytically superior, no further comment will be made on oblique rotational methods. Rather the focus will be on orthogonal approaches.

In practice the objective of all methods of rotation is to simplify the rows and/or the columns of the factor matrix to facilitate interpretation. By simplifying the rows we mean making as many values in each row as close to zero as possible. By simplifying the columns we mean making as many values in each column as close to zero as possible. Three major orthogonal approaches have been developed. They are QUARTIMAX, VARIMAX and EQUIMAX. The ultimate goal of a QUARTIMAX rotation is to simplify the rows of a factor matrix. That is, it focuses on rotating the initial factor so a variable loads high on one factor and as low as possible on all other factors. In contrast to QUARTIMAX, the VARIMAX criterion centers on simplifying the columns of the factor matrix. Note that in QUARTIMAX approaches many variables can load high or near high on the same factor because the technique centers on simplifying the rows. With the VARIMAX rotational approach, the maximum possible simplification is reached if there are only ones and zeros in a single column. The EQUIMAX approach is a compromise between the QUARTIMAX and VARIMAX criteria. Rather than concentrating either on simplification of the rows or on simplification of the columns, it tries to accomplish some of each; thus the name EQUIMAX is used for this approach.

No specific rules have been developed to guide the analyst in selecting a particular orthogonal rotational technique. In most instances the analyst will simply utilize the rotational technique which is a standard output of the computer program used. Most programs have only a single rotational option and it is usually VARIMAX. Thus, the VARIMAX method of rotation is the most widely utilized. However, there is no compelling analytical reason to favor one rotational method over another. Whenever possible, the choice should be made on the basis of the particular needs of a given research problem.

Criteria for the Number of Factors to be Extracted

How do we decide on how many factors to extract? When a large set of variables is factored, the analysis will extract the largest and best combinations of variables first, and then proceed to smaller, less understandable combinations. In deciding when to stop factoring (that is, how many factors to extract), the analyst generally will begin with some predetermined criteria such as the a priori criterion, or the latent root criterion, to arrive at a specific number of factors to extract (these two techniques will be discussed in more detail later). After the initial solution has been derived, the analyst will then make several additional trial rotations—usually one less factor than the initial number, and two or three more factors than were initially derived. Then, on the basis of information contained in the results of these several trial analyses, the factor matrices will be examined and the best representation of the data will be used to assist in determining the number of factors to extract. To use a microscope analogy, choosing the number of factors to be interpreted is something like focusing. Too high or too low an adjustment will obscure a structure that is obvious when the adjustment is just right. Therefore, by examining a number of different factor structures derived through several trial rotations, the analyst can compare and contrast to arrive at the best representation of the data.

An exact quantitative basis for deciding the number of factors to extract has not been developed. However, the following stopping criteria for the number of factors to extract are currently being utilized.

(1) The most commonly used technique is referred to as the **latent root criterion.** This rule is very simple to apply. But it does differ depending on whether the analyst has chosen either component analysis or common factor analysis as the basic model. Recall that in component analysis ones are inserted in the diagonal of the correlation matrix and the entire variance is considered in the analysis. In component analysis only the factors having latent roots (eigenvalues) greater than one are considered significant; all factors with latent roots less than one are considered insignificant and disregarded.

Many factor analysts utilize only the eigenvalue one criteria. However, when the common factor model is selected, the eigenvalue one criteria should be adjusted slightly downward. With the common factor model, the eigenvalue cutoff level should be lower and approximate either the estimate for the common variance of the set of variables, or the average of the communality estimates for all variables.

The rationale for the eigenvalue criteria is that any individual factor should account for at least the variance of a single variable if it is to be retained for interpretation. The eigenvalue approach is probably most reliable when the number of variables is between 20 and 50. In instances where the number of variables is less than 20 there is somewhat of a tendency for this method to extract a conservative number of factors. When more than 50 variables are involved, however, it is not uncommon for too many factors to be extracted.

(2) The **a priori criterion.** This is a simplistic, yet reasonable criterion under certain circumstances. When applying the a priori criterion the analyst already knows how many factors to extract before undertaking the factor analysis. The analyst simply instructs the computer to stop the analysis when the desired number of factors has been extracted. This approach is useful if the analyst is testing a theory or hypothesis about the number of factors to be extracted. It also can be justified in instances where the analyst is attempting to replicate another researcher's work and extract exactly the same number of factors that was previously found.

(3) The **percentage of variance criterion** is another approach. Using this approach, the cumulative percentages of the variance extracted by successive factors is the criterion. No absolute cutting line has been adopted for all data. However, in the hard sciences the factoring procedure usually should not be stopped until the extracted factors account for at least 95 percent of the variance, or the last factor accounts for only a small portion (less than 5 percent) of the variance. In contrast, in the social sciences where information is often less precise, it is not uncommon for the analyst to consider a solution which accounts for 60 percent of the total variance (and in some instances even less) as a satisfactory solution.

(4) The **scree test criterion.** Recall that with the component analysis factor model the later factors extracted contain both common and unique variance. While all factors contain at least some unique variance, the proportion of unique variance in later factors is substantially higher than in earlier factors. The scree tail test is an approach used to identify the optimum number of factors which can be extracted before the amount of unique variance begins to dominate the common variance structure [2]. The scree test is derived by plotting the latent roots against the number of factors in their order of extraction, and the shape of the resulting curve is used to evaluate the cutoff point. Figure 6.5 plots the first 18 factors extracted in a recent study by the authors. Starting with the first factor, the plot slopes steeply down initially and then slowly becomes an approximately horizontal line. The point at which the curve first begins to straighten out is considered to be the maximum number of factors to extract. In the present case the first ten factors would qualify. Beyond ten factors, too large a proportion of unique variance would be in-

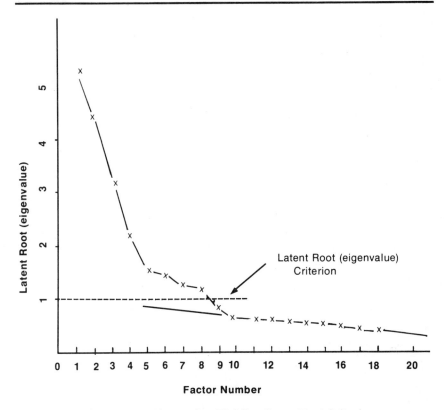

Figure 6.5—Eigenvalue Plot For Scree Test Criterion

cluded thus these factors would not be acceptable. Note that in using the latent root criterion only eight factors would have been considered. In contrast, using the scree test provides us with two more factors. As a general rule, the scree tail test will result in at least one and sometimes two or three more factors being considered as significant than will the latent root criterion [2].

In practice, most factor analysts seldom use a single criterion for selecting how many factors to extract. Instead, they initially use a criterion such as the latent root as a guideline for the first rotation. Then, several additional trial rotations are undertaken, and by considering the initial criterion and comparing the factor interpretations for several different trial rotations, the analyst can select the number of factors to extract based upon the initial criterion and the factor structure which best represents the underlying relationship of the variables. In short, the ability to assign some meaning to the factors, or to interpret the nature of the variables, becomes an extremely important consideration in determining the number of factors to extract.

Criteria for the Significance of Factor Loadings

In interpreting factors a decision must be made regarding which factor loadings are worth considering.

(1) The first suggestion is not based on any mathematical proposition except that it represents approximately 10 percent of the variance of a particular variable. It is a rule of thumb which has been used frequently by factor analysts as a means of making a preliminary examination of the factor matrix. In short, factor loadings greater than ± .30 are considered significant. Loadings ± .40 are considered more important, and if the loadings are ± .50 or greater they are considered very significant. Thus, the larger the absolute size of the factor loading the more significant the loading is in interpreting the factor matrix. These guidelines are considered useful when the sample size is 50 or larger. This approach may appear too simplistic, yet compared with other criteria it is a quite rigorous and acceptable approach.

(2) As pointed out previously, a factor loading represents the correlation between an original variable and its respective factor. In determining a significance level for interpretation of loadings, an approach could be used which is similar to that of interpreting correlation coefficients [3]. Specifically, loadings of at least ± .19 and ± .26 are recommended for the five percent and one percent levels respectively when the sample size is 100. When the sample size is 200, ± .14 and ± .18 are recommended for the five percent and one percent levels of significance. Finally when the sample size is at least 300, loadings of ± .11 and ± .15 are recommended for the five percent and one percent levels respectively. Since it is difficult to assess the amount of error involved in factor analytic studies it is probably safer to adopt the one percent level as the criterion for significance.

(3) A disadvantage of methods 1 and 2 is that the number of variables being analyzed or the specific factor being examined are not considered. It has been shown that as the analyst moves from the first factor to later factors the acceptable level for a loading to be judged significant should increase. The fact that unique variance and error variance begin to enter in later factors means that some adjustment upward in the level of significance should be included [3].

The number of variables being analyzed is also important in deciding which loadings are significant. As the number of variables being analyzed increases the acceptable level for considering a loading significant decreases. Adjustment for the number of variables is particularly true as you move from the first factor extracted to later factors. Specifically when the sample size is 50 and the desired significance level is .05, the following guidelines are applicable: (1) a significant loading on the fifth factor with 20 variables would be ± .292 whereas with 50 variables a significant loading would be ± .267, and (2) a significant loading on the tenth factor with 20 variables would be ± .353, but with 50 variables it would only be ± .274. Similar guidelines can be given when the sample size is 100 and the significance level is .05: (1) a significant loading on the fifth factor with 20 variables would be ± .216, but with 50 variables it would drop to ± .202, and (2) a significant loading on the tenth factor with 20 variables would be ± .261, but only ± .214 with 50 variables [3].

To summarize the criteria for the significance of factor loadings, the following guidelines can be stated: (1) the larger the sample size the smaller the loading to be considered significant; (2) the larger the number of variables being analyzed the smaller the loading to be considered significant, and (3) the larger the number of factors, the larger the size of the loading on later factors to be considered significant for interpretation.

Interpreting a Factor Matrix

Interpreting the complex interrelationships represented in a factor matrix is no simple matter. By following the procedure outlined in the following paragraphs, however, the factor interpretation procedure can be simplified considerably.

(1) Examine the factor matrix. Each of the columns of numbers represents a separate factor. The columns of numbers are the factor loadings for each variable on each factor. For identification purposes, the computer printout will usually identify the factors from left to right by the numbers 1, 2, 3, 4 and so forth. It also will identify the variables by number from top to bottom. To further facilitate interpretation the analyst should write out the name of each variable in the left margin beside the variable numbers.

(2) To begin the interpretation the analyst should start with the first variable on the first factor and move horizontally from left to right looking for the **highest loading** for that variable on any factor. When the highest loading (largest absolute factor loading) is identified, if it is significant the analyst should underline it. The analyst should then go to the second variable and again moving from left to right horizontally look for the highest loading for that variable on any factor and underline it. Continue this procedure for each variable

until all variables have been underlined once for their highest loading on a factor. Recall that for sample sizes of less than 100 the lowest factor loading to be considered as significant would in most instances be ± .30.

It should be noted that the process of underlining only the single highest loading as significant for each variable is an ideal that the analyst should strive for, but can seldom achieve. When each variable has only one loading on one factor that is considered significant, the interpretation of the meaning of each factor is simplified considerably. In practice, many variables will have several moderately sized loadings, all of which are significant, and the job of interpreting the factors is much more difficult. This is because a variable with several significant loadings must be considered in interpreting (labeling) all the factors on which it has a significant loading. Since most factor solutions do not result in a simple structure solution (a single high loading for each variable on only one factor), the analyst will, after underlining the highest loading for a variable, continue to evaluate the factor matrix by underlining all significant loadings for a variable on all the factors. Ultimately, the analyst tries to minimize the number of significant loadings on each row of the factor matrix (that is, the loadings associated with one variable) and to maximize the number of loadings with negligible values.

(3) Once all the variables have been underlined on their respective factors, then the analyst should examine the factor matrix to identify variables which have not been underlined and therefore do not "load" on any factor. If there are variables which do not load on any factor then the analyst has two options: (1) interpret the solution as it is and simply ignore those variables without a significant loading, or (2) critically evaluate each of the variables that do not load significantly on any factor. This evaluation would be in terms of the variable's overall contribution to the research as well as its communality index. If the variable(s) is of minor importance to the study's objective and/or has a low communality index, the analyst may decide to eliminate the variable or variables and derive a new factor solution with the "non-loading" variables eliminated.

(4) When a factor solution has been obtained in which all significant variables are loading on a factor, then the analyst will attempt to assign some meaning to the pattern of factor loadings. Variables with higher loadings are considered more important in this stage of the factor interpretation. They influence the name or label selected to represent a factor to a greater extent. Thus, the analyst will examine all the underlined variables for a particular factor and, placing greater

emphasis on those variables with higher loadings, will attempt to assign a name or label to a factor which accurately reflects to the greatest extent possible what the several variables loading on that factor represent. It is important to note that this label is not derived or assigned by the factor analysis computer program, but rather is intuitively developed by the factor analyst based upon its appropriateness for representing the underlying dimensions of a particular factor. This procedure is followed for each of the extracted factors. The final result will be a name or label which as accurately as possible represents each of the derived factors.

In some instances it is not possible to assign a name to each of the factors. When such a situation is encountered the analyst may wish to use the label "undefined" to represent a particular factor or factors derived by that solution. In such cases the analyst would interpret only those factors which are meaningful and would disregard those undefined or less meaningful factors. It is important to note, however, that in describing the factor solution the factor analyst would indicate that these factors were derived but were undefinable, and only those factors representing meaningful relationships were interpreted.

AN ILLUSTRATIVE EXAMPLE

In the preceding pages the major empirical questions concerning the application of factor analysis have been covered. To further clarify these topics, an illustrative example of the application of factor analysis is presented based upon data from the data bank presented in Chapter One of this text. The first six variables from the data bank were selected for conducting the factor analysis. Our focus here will be to take the reader through a step-by-step application and interpretation of a component analysis and a common factor analysis of the same six variables. The rotational approach will be orthogonal and VARIMAX for both factor models. First we shall consider component analysis and then we will look at common factor analysis.

Component Analysis

As noted earlier, factor analysis procedures are based upon the initial computation of a complete table of intercorrelations among the variables (correlation matrix). This correlation matrix is then transformed to obtain a factor matrix. The correlation matrix provides an initial indication of the relationships among the variables. However, interpretation of the correlation matrix still involves the examination

TABLE 6.3
COMPONENT ANALYSIS CORRELATION MATRIX

Variable	(1)	(2)	(3)	(4)	(5)	(6)
		Correlations Among Variables				
X_1 Self-Esteem	1.00	−.38	.51	.02	.61	.04
X_2 Locus of Control		1.00	−.50	.23	.48	.14
X_3 Alienation			1.00	−.13	.06	−.05
X_4 Social Responsibility				1.00	.24	.77
X_5 Machiavellianism					1.00	.16
X_6 Political Opinion						1.00

of numerous relationships and can be extremely difficult. For example, even 20 variables will yield a correlation matrix containing 210 separate entries. Since such a table of numbers is frequently too large to grasp and interpret effectively, the question arises if it might be possible to develop an even more condensed arrangement that will represent the underlying order in the data better. In many instances the answer to this question will be yes. Through the application of factor analytic techniques it is frequently possible to reduce the correlation matrix to a smaller set of relationships—the factor matrix. Of course, this assumes that a certain degree of underlying order exists in the data being analyzed.

Table 6.3 shows the correlation matrix for the six variables drawn from the data bank. From inspection of the correlation matrix we can see that six variables are related at the .38 level or above (underlined). But it is difficult to derive a complete and clear understanding of the relationships. From a factor analysis of these variables it should be possible to derive a clearer picture of the underlying relationships of the variables.

The result of the first stage in the computation of factors is shown in Table 6.4—the unrotated component analysis factor matrix. To begin the analysis let's explain the numbers included in the table. Four columns of numbers are shown. The first three columns of numbers are the results for the three factors which are extracted. The fourth column of numbers provides summary statistics covering all the factors in this particular factor solution. The matrix of factor loadings consists of the three columns of numbers beside the six variables. The numbers at the bottom of each of the three columns are the *column sum of squared factor loadings* and the percent of trace. The column sum of squared factor loadings (eigenvalue) shown

TABLE 6.4
UNROTATED COMPONENT ANALYSIS FACTOR MATRIX

Variables	Factors (1)	(2)	(3)	Communality
X_1 Self-Esteem	−.08	.93	.20	.91
X_2 Locus of Control	.69	−.38	.53	.90
X_3 Alienation	−.42	.71	−.19	.72
X_4 Social Responsibility	.81	.18	−.45	.89
X_5 Machiavellianism	.51	.57	.64	.99
X_6 Political Opinion	.73	.22	−.55	.88
Sum of Squares (eigenvalue)	2.13	1.90	1.27	5.29
Percent of Trace[1]	35.50	31.67	21.17	88.34

[1]Trace = 6.00

at the bottom of each of the columns of factor loadings indicates the relative importance of each factor in accounting for the variance associated with the set of variables being analyzed. Note the sums of squares for factors one, two, and three are 2.13, 1.90, and 1.27, respectively. As expected, the unrotated factor solution has extracted the factors in the order of their importance with one accounting for the most variance, two slightly less and the third factor accounting for the least amount of variance. At the far right hand side of the row of sums of squares is the number 5.29 which represents the total sum of squares. The total sum of squared factor loadings is obtained by adding the individual sums of squares for each of the factors. It represents the total amount of variance extracted by the factor solution.

The **percentage of trace** for each of the three factors is also shown at the bottom of the table. The percentages of trace for factors one, two, and three are 35.50, 31.67, and 21.17, respectively. The percent of trace is obtained by dividing each factor's sum of squares by the trace for the set of variables being analyzed. For example, if the sum of squares of 2.13 for factor one is divided by the trace of 6.0 the result will be the percent of trace, or 35.50 percent for factor one. By adding the percentages of trace for each of the three factors together, we obtain the total percent of trace extracted for the factor solution. The total percent of trace can be used as an index to determine how well a particular factor solution accounts for what all the variables together represent. If the variables are all very different from each other this index will be low. If the variables fall into one or more highly redundant or related groups, and if the extracted factors account for all the groups, the index will approach 100. The index for the present solution shows that 88.34 percent of the total variance is represented by the information contained in the factor matrix. Therefore the index for this solution is high and the variables are in fact highly related to each other.

The **row sum of squared factor loadings** is shown at the far right side of the table. These figures, referred to in the table as **communalities,** show the amount of variance in a variable that is accounted for by the three factors taken together. The size of the communality is a useful index for assessing how much variance in a particular variable is accounted for by the factor solution. Large communalities indicate that a large amount of the variance in a variable has been extracted by the factor solution. Small communalities show that a substantial portion of the variance in a variable is unaccounted for by the factors. For instance, the communality figure of .72 for variable X_3 indicates it has less in common with the other variables included in the analysis than does variable X_5 which has a communality of .99.

Having defined the various elements of the unrotated factor matrix, let's examine the factor loading patterns. As anticipated, the first factor accounts for the largest amount of variance and is a general factor with every variable except X_1 loading significantly. Also, based upon the sample size, the number of variables and the factor number, variable X_2 loads on all three factors significantly. Variable X_3 loads on factors one and two, variable X_4 loads on factors one and three, variable X_5 loads on all three factors, and variable X_6 loads on factors one and three. Based on this factor loading pattern, interpretation would be extremely difficult and theoretically less meaningful. Therefore the analyst should proceed to rotate the factor matrix to redistribute the variance from the earlier factors to the later factors. This should result in a simpler and theoretically more meaningful factor pattern.

The VARIMAX rotated component analysis factor matrix is shown in Table 6.5. Note that the total amount of variance extracted is virtually the same in the rotated solution as it was in the unrotated one—88.83 percent (the difference—88.34 versus 88.83—is due to computational rounding). Two major differences are obvious, however. The variance has been redistributed so that the factor loading pattern is different, and the percentage of variance for each of the factors is different because the factors are not extracted in their order of im-

TABLE 6.5
VARIMAX ROTATED COMPONENT ANALYSIS FACTOR MATRIX

Variables	Factors (1)	(2)	(3)	Communality
X_1 Self-Esteem	.02	.65	.71	.92
X_2 Locus of Control	.11	− .87	.37	.01
X_3 Alienation	− .05	.83	.18	.72
X_4 Social Responsibility	.93	− .12	.11	.89
X_5 Machiavellianism	.13	− .13	.98	.99
X_6 Political Opinion	.95	− .00	.04	.90
Sum of Squares (eigenvalue)	1.79	1.90	1.64	5.33
Percent of Variance	29.83	31.67	27.33	88.83

portance based on the amount of variance extracted. Specifically, in the VARIMAX rotated factor solution the first factor accounts for 29.83 percent of the variance, but the second factor accounts for 31.67 percent of the variance. The third factor accounts for 27.33 percent of the variance. Also recall that in the unrotated factor solution all variables except X_1 loaded significantly on the first factor. In the rotated factor solution, however, variables X_4 and X_6 load significantly on factor one, variables X_1, X_2 and X_3 load significantly on factor two, and variables X_1 and X_5 load significantly on factor three. The only variable which loads significantly on two factors is X_1 which loads on factors two and three. To some extent the analyst may consider variable X_2 as also loading significantly on two factors—two and three. The loading of .37 for variable X_2 on factor three could be considered significant. But since the difference between .37 and the next highest loading of .65 for variable X_1 on factor two is so great, the analyst probably would consider this differential as too large and the loading for variable X_2 would not be judged as significant. It should be apparent that factor interpretation has been simplified considerably by rotating the factor matrix.

Common Factor Analysis

Common factor analysis is one of the two major factor analytic models that will be discussed here. The difference between component analysis and common factor analysis is that the latter considers only the common variance associated with a set of variables. This is

TABLE 6.6
COMMON FACTOR ANALYSIS CORRELATION MATRIX

Variable	Correlations Among Variables					
	(1)	(2)	(3)	(4)	(5)	(6)
X₁ Self-Esteem	.97	−.35	.51	.04	.59	.03
X₂ Locus of Control		.97	−.48	.25	.46	.13
X₃ Alienation			.38	−.08	.05	−.07
X₄ Social Responsibility				.63	.22	.48
X₅ Machiavellianism					.98	.17
X₆ Political Opinion						.61

accomplished by factoring a "reduced" correlation matrix with communalities in the diagonal instead of unities. Comparison of the numbers on the diagonal of the common factor correlation matrix (Table 6.6) with the numbers shown on the diagonal of the component analysis correlation matrix (Table 6.3) will demonstrate the difference between the correlation matrix for the two factor models. The numbers on the diagonal of the component analysis correlation matrix are all ones, whereas for the common factor model all the numbers on the diagonal are less than one.

Examination of the sizes of the communalities for each of the variables will suggest whether the variables' loading pattern in the common factor solution will differ from the component analysis solution. Specifically, the communalities for variables X_1, X_2 and X_5 are all very high. Therefore the factor loading pattern for these variables in the common factor model should not differ substantially from what it was in the component analysis. In contrast, the communalities for variables X_4, X_6 and particularly X_3 are much lower than with the component analysis correlation matrix. This suggests that the loading pattern for these three variables will probably be different than it was with the component factor model. This difference may result in the variables loading on a different factor or it may simply mean that they load on the same factor as with the component model, but the size of the loading may differ since their communality is lower.

Turning next to the VARIMAX rotated common factor analysis factor matrix (Table 6.7), let's examine how it compares with the component analysis rotated factor matrix. The information provided for the common factor solution is similar to that for the component analysis solution. Sums of squares, percent of variance, communalities, total sums of squares, and total variance extracted are all provided just as with the component analysis solution. The information which

TABLE 6.7
VARIMAX ROTATED COMMON FACTOR ANALYSIS FACTOR MATRIX

Variables	Factors (1)	(2)	(3)	Communality
X_1 Self-Esteem	.65	.74	.04	.97
X_2 Locus of Control	.42	−.87	.13	.96
X_3 Alienation	.10	.60	−.06	.37
X_4 Social Responsibility	.12	−.10	.73	.56
X_5 Machiavellianism	.97	−.05	.16	.97
X_6 Political Opinion	.04	−.03	.74	.55
Sum of Squares (eigenvalue)	1.57	1.68	1.13	4.38
Percent of Trace[1]	34.58	37.00	24.89	96.47
Percent of Variance	26.17	28.00	18.83	73.00

[1] Trace = 4.54 = 75.6 percent of total variation is common variance.

differs is the row referred to as Percent of Trace. Recall that with component analysis the trace is equal to the number of variables (based on the assumption that the variance in each variable is equal to one). In contrast, the trace for a common factor analysis solution is equal to the sum of the communalities on the diagonal of the correlation matrix. For the present example the sum of the communalities (trace) is 4.54. The percent of trace is included to demonstrate the relative importance of each of the factors in accounting for the variance included in the reduced common factor solution. For example, by dividing the sum of squares for factor one (1.57) by the trace (4.54), we obtain the percent of trace for factor one—34.58 percent. By summing the percent of trace for the factors we obtain the total percent of trace for the factor solution. For the present example it is 96.47 percent. This percent is useful because it tells the analyst how much of the common variance (trace) is accounted for by the factor solution.

Another percentage which may be useful to the analyst is the percent of total variation which is common variance. As can be noted, the amount of variance included in the reduced common factor solution is 4.54 as opposed to 6.0. Thus, by dividing the amount of common variance (4.54) by the amount of total variance (6.0), we obtain the percent of total variation that is common variance. The result of this computation reveals that 75.6 percent of the total variation is common.

Comparison of the information provided in the common factor analysis factor matrix and the component analysis factor matrix shows several differences. The factor loading pattern is similar in that variables X_1 and X_5 load together, variables X_1, X_2 and X_3 load together, and variables X_4 and X_6 load together. The primary difference, is that with the component analysis factor matrix the factor with variables X_4 and X_6 loading was factor number one, whereas with the common factor analysis variable X_4 and X_6 loaded on factor three. Similarly, variables X_1 and X_5 loaded on factor three with the component analysis solution, whereas these same variables loaded on factor one with the common factor solution. The logic for this reversal in the order of extraction for the factors is based upon the communalities for the respective variables. Specifically, note that the communalities for variables X_4 and X_6 are relatively low. Therefore, when the factor solution is based only on the common variance, as with a common factor solution, then the order of extraction would dictate that factors with variables exhibiting lower communalities would be extracted later than they would be under a component analysis when the total variance is included in the correlation matrix.

Naming of Factors

When a satisfactory factor solution has been derived, the analyst usually will attempt to assign some meaning to it. The process involves substantive interpretation of the pattern of factor loadings for the variables, including their signs, in an effort to name each of the factors. Before interpretation, a minimum acceptable level of significance for a factor loading must be selected. All significant factor loadings typically are used in the interpretation process. But variables with higher loadings will influence the name or label selected to represent a factor to a greater extent. The signs are interpreted just as with any other correlation coefficients. On each factor like signs mean the variables are positively related and opposite signs mean the variables are negatively related. In orthogonal solutions the factors are independent of each other. Therefore, the signs for a factor loading relate only to the factor which they appear on and not to other factors in the solution.

Let's look at the results shown in Table 6.7 to illustrate this procedure. Our factor solution was derived from a common factor VARIMAX rotation of the six social-psychological variables. Our cutoff point for this solution is all loadings $\pm$ .60 or above (underlined in Table 6.7). This relatively high cutoff was possible because many high loadings were obtained. Variable X_2—locus of control—has a loading of .42 on factor one and could be considered significant. But it is loading substantially below all the other variables considered significant (.42 versus .60 or higher), is only about half the size of its loading on factor two ($-.87$), and to include it would violate the guidelines for simple structure factor solutions (only one loading on any factor for each variable). Thus, it was not considered significant.

Substantive interpretation is based on the significant higher loadings. Factor one has two significant loadings, factor two has three significant loadings, and factor three has two significant loadings (significant loadings are underlined). Looking at factor one we see that Variables X_1—Self-Esteem and X_5—Machiavellianism are positively related to each other. This suggests that persons in our sample who are high in self-esteem are also high in Machiavellianism. A possible name for this factor on the loading pattern is "Indiscriminant High Confidence Achiever." But perhaps you may want to assign your own label.

Turning next to factor two we note that Variables X_1—Self-Esteem and X_3—Alienation are positively related to each other, and negatively related to Variable X_2—Locus of Control. From the description of our data bank in Chapter One we know the following scoring procedures for the variables: X_1—Self-Esteem = high scores indicate higher self-estem, X_2—Locus of Control = high scores indicate external locus of control, and X_3—Alienation = high scores indicate a person who is not alienated. From this scoring procedure and the signs of the variable loadings, we can interpret this factor as an individual who is high in self-esteem, exhibits an internal locus of control, and is not alienated (the opposite interpretation for all three variables is also possible). A possible label for this factor is "Self-Confident Achiever." But again you may wish to develop your own name to represent the factor.

The process of naming factors has been demonstrated. You will note that it is not very scientific and is based on the subjective opinion of the analyst. Different analysts will no doubt assign different names to the same results because of the difference in their background and training. For this reason the process of labeling factors is subject to considerable criticism. But if a logical name can be assigned that represents the underlying nature of the factors, it usually facilitates the presentation and understanding of the factor solution and therefore is a justifiable procedure.

How to Select Surrogate Variables for Subsequent Analysis

If the researcher's objective is to identify appropriate variables for subsequent application with other statistical techniques (purpose three), then the researcher would examine the factor matrix and select the variable with the highest factor loading as a surrogate representative for a particular factor dimension. If there is one factor loading for a variable which is substantially higher than all other factor loadings, then the variable with the obviously higher loading would be selected for subsequent analysis to represent that factor. In some instances, the selection process is more difficult because two or more variables have loadings which are significant and fairly close to each other. In such cases the analyst would have to critically examine the several factor loadings which are of approximately the same size and select only one as a representative of a particular dimension. This decision would be based on the researcher's a priori knowledge of theory which would suggest that a particular variable would be more logically representative of the dimension which has been identified. Also, the analyst may have knowledge which suggests that the raw data for a variable which is loading slightly lower is in fact more reliable than the raw data for the highest loading variable. In such cases the analyst may choose the variable which is loading slightly lower as the variable to represent a particular factor.

Let's examine the data provided in Table 6.5 to clarify the procedure for selecting surrogate variables. First, recall that surrogate variables would be selected only when the rotation is orthogonal. This is because when the analyst is interested in using surrogate variables in subsequent analyses, he would want to observe to the extent possible the assumption that the independent variables should be uncorrelated with each other. Thus an orthogonal solution would be selected instead of an oblique one. Focusing on the factor loadings for factor one we see that the loading for variable X_4 is .93 and for variable X_6 is .95. The selection of a surrogate is difficult in cases like this because the sizes of the loadings are so close. However, if the analyst has no a priori evidence to suggest that the reliability or validity of the raw data for one of variables is better than for the other, and if neither would be more theoretically meaningful for the factor interpretation, then the analyst would select variable X_6 as the surrogate variable. In contrast, the loadings for factor three are .98 for variable X_5 and .71 for variable X_1. Therefore, because of its substantially higher loading the analyst would select variable X_5 as the surrogate variable to represent factor three in a subsequent analysis.

How to Use Factor Scores

When the analyst is interested in creating an entirely new set of a smaller number of composite variables to replace either in part or completely the original set of variables, then the analyst would compute factor scores (purpose four). Factor scores are composite measures for each factor representing each subject. The original raw data measurements and the factor analytic results are utilized to compute factor scores for each individual. Using our data bank example, each individual would have had six raw data measurements representing each of the original six variables. After computation of factor scores to represent the factor solution, each individual would be represented by only three composite measures rather than the original six measures. These three composite measures or factor scores would represent each of the three factors which were derived in the factor solution. Conceptually speaking, the factor score represents the degree to which each individual scores high on the group of items that load high on a factor. Thus, an individual who scores high on the several variables that have heavy loadings for a factor surely will obtain a high factor score on that factor. The factor score, therefore, shows that an individual possesses a particular characteristic represented by the factor to a high degree. Most factor analysis computer programs compute scores for each respondent on each factor to be utilized in subsequent analysis. The analyst would merely have to select the factor score option and these scores would either be printed out or punched out for use by the analyst.

SUMMARY

The multivariate statistical technique of factor analysis has been presented in broad conceptual terms. Basic guidelines for interpreting the results were included to further clarify the methodological concepts. An example application of factor analysis was presented based upon the first chapter data bank.

Factor analysis can be a highly useful and powerful multivariate statistical technique for effectively extracting information from large data bases. Factor analysis helps the investigator make sense of large bodies of interrelated data. When it works well, it points to interesting relationships that might not have been obvious from examination of the raw data alone, or even a correlation matrix. Potential applications of factor analytic techniques to problem-solving and decision-making in business research are numerous. The use of these techniques will continue to grow as increased familiarity with the procedures is gained by academicians and practitioners.

Factor analysis is a much more complex and lengthy subject than might be indicated by this brief exposition. Following are four of the most frequently cited limitations. First, there are many techniques for performing factor analyses. Controversy exists over which technique is best. Second, the subjective aspects of factor analysis (deciding how many factors to extract, which technique should be used to rotate the factor axes, which factor loadings are significant), are all subject to many differences in opinion. Third, the computational labor in conducting factor analysis and any other multivariate technique with large data bases necessitates the use of computers. With the rapid spread of computers, this particular limitation has diminished. Fourth, the problem of reliability is very real. Like any other statistical procedure a factor analysis starts with a set of imperfect data. When the data changes because of changes in the sample, the data gathering process, or the numerous kinds of measurement errors, the results of the analysis will change also. The results of any single analysis are therefore less than perfectly dependable. This problem is especially critical because the results of a single factor analytic solution frequently look plausible. It is important to emphasize that plausibility is no guarantee of validity or even stability.

END OF CHAPTER QUESTIONS

1. What are three problem situations in which factor analysis is the appropriate multivariate statistical technique to apply?
2. What is the difference between an orthogonal factor rotation and an oblique one? When would the application of each approach be most appropriate? ·
3. What guidelines can you use to determine the number of factors to extract? Explain each briefly.
4. How do you use a factor loading matrix to interpret the meaning of factors?
5. How and when should you use factor scores in conjunction with other multivariate statistical techniques?

REFERENCES

1. Anderson, T. W., Introduction to Multivariate Statistical Analysis, New York: John Wiley & Sons, Inc., 1958.
2. Cattell, R. B., "The Scree Test for the Number of Factors," Multivariate Behavioral Research, Vol. 1 (April, 1966): pp. 245-76.
3. Child, D., The Essentials of Factor Analysis, New York: Holt, Rinehart and Winston, Inc., 1970.
4. Cooley, W. W., and P. R. Lohnes, Multivariate Data Analysis. New York: John Wiley & Sons, Inc., 1971.
5. Dixon, W. J., Biomedical Computer Programs. Los Angeles: University of California Press, 1967.
6. Green, Paul E., and Donald S. Tull, Research for Marketing Decisions Englewood Cliffs: Prentice-Hall, Inc., 1975.

7. Harris, R. J., *A Primer of Multivariate Statistics,* New York: Academic Press, 1975.
8. Nie, N., C. Hull, J. Jenkins, K. Steinbrenner, and D. Bent, *Statistical Package for the Social Sciences,* New York: McGraw-Hill, 1975.
9. Overall, J. E., and J. Klett, *Applied Multivariate Analysis.* New York: McGraw-Hill, Inc., 1972.
10. Veldman, D., *Fortran Programming for the Behavioral Sciences.* New York: Holt, Rinehart, Winston, Inc., 1967.
11. Wells, W. D., and J. Sheth, "Factor Analysis in Marketing Research," from *Handbook of Market Research,* Robert Ferber (ed.), New York: McGraw-Hill, 1971.

SELECTED READINGS

CONSUMER ATTITUDES TOWARD GOVERNMENT INTERVENTION AND MARKETING PRACTICES: A FACTOR ANALYTIC APPROACH

WILLIAM D. PERREAULT, JR. and HIRAM C. BARKSDALE†

Introduction

Many of the research problems in business are concerned with the interrelationships of multiple variables, and thus it is a natural evolution that multivariate statistical methods have increasingly been used to reveal the underlying processes reflected in business data. Factor analysis is one such technique that has proved particularly useful as a means of reducing the complexity of business data, and bringing parsimony to research efforts. There are many different ways in which factor analysis has been used in business research (see, for example [2, 10, 15, 19, 24, 25]), yet frequently in published research little detail is given concerning the specific judgments that were involved in factor analysis applications.

Thus, the purpose of this article is to discuss and illustrate the use of factor analysis in one of its common applications: the development of (multi-item) scales which are to be used in subsequent analysis. Specifically, results of a factor analysis of scale statements are presented and discussed in detail to further clarify the nature of the analysis and its use. The paper consists of four sections. In the

† William D. Perreault, Jr. is on the Marketing faculty of the Graduate School of Business Administration at the University of North Carolina, Chapel Hill. Hiram C. Barksdale is on the Marketing faculty of the College of Business at the University of Georgia, Athens.
* Prepared especially for this text.

first, we briefly describe the research problem to which factor analysis is applied; in the second, we present the results of the analysis, and discuss the interpretation of the resulting statistics. In the third section, we use the scales developed by the factor analysis as input to further analysis. In conclusion, we overview several emerging issues concerning the use of factor analysis as applied in business research.

AN EXAMPLE FROM CONSUMERISM

Recently, practitioners and scholars of the business community have paid increased attention to the forces of consumerism. Business executives, public officials, and researchers have watched the consumerist wave and attempted to determine its implications for the workings of the market place [1, 11, 14, 18]. Many companies have taken explicit steps to be more sensitive to consumer satisfaction with and attitudes toward business practices [5], and public officials have made policy initiatives to develop regulations which better serve or protect consumer interests.

One important question, then, is whether or not consumers' attitudes toward marketing practices and government intervention have changed during the period in which these efforts by business and government have evolved. It is this basic question which motivates the analysis presented in this paper.

One way to evaluate changes in consumer attitudes is to measure those attitudes at different points in time, and then compare what has happened across the temporal dimension. This is the approach to be used here. This study draws upon data collected in four national mail surveys: the first survey was conducted in 1971, and then the study was replicated in 1973, 1975, and 1977. Initial results of this research program have previously been published, and these earlier reports [3, 4] provide details concerning the sampling procedures, size and composition of the samples, and nature of the studies. However, this is the first report of the 1977 survey, and thus it adds meaningfully to the longitudinal dimension of the study. In the 1977 survey, questionnaires were sent to 1200 consumers, and 641 questionnaires were completed and returned. The factor analysis reported here is based on data from this survey.

Specifically, the concern here is to develop good composite (scale) measures reflecting consumer attitudes toward (1) government intervention and (2) the consumer orientation of business. In the survey questionnaires, there were a subset of statements dealing with these topics. Each of these statements suggests an attitude about business or government. Responding consumers indicated their level of agreement with each of the statements, and the responses were scored from 1 (strongly agree) to 5 (strongly disagree). The statements are displayed in Table 1, along with the means and standard deviations for the 1977 survey responses. The reader may note from studying Table 1 that several of the statements (number 3 and number 6) were worded in the negative, and thus they were reverse scored for analysis. Also, note that the statement pool is grouped into two sections: those that seemed to reflect various attitudes toward government intervention (or non-intervention) in the market place, and those that reflect consumer attitudes toward business practices, and in particular the extent to which business is "consumer oriented." It should be emphasized, however, that the grouping of the items implied in Table 1 is based on their *apparent* content, and not on an empirical evaluation or verification that the groupings are appropriate for scale development.

Factor analysis is a helpful input to such an evaluation. It can help to reveal the "underlying attitudes" (dimensions) which tend to be associated with (subgroups of) the statements, and to identify the grouping of statements which are most closely associated or aligned with those dimensions. With such information, it is possible to develop good summary scale measures based on multiple items. Multi-item scales developed in this fashion are desirable because they tend to be more *reliable* than measures based on the original, individual items [21].

FACTOR ANALYSIS RESULTS

Inter-Statement Correlations

The first step in the analysis is the computation of the product moment correlation matrix which summarizes the linear associations among the responses to each pair of statements (items). Intuitively, if the first five statements from Table 1 do in fact reflect a common underlying attitude toward government intervention, we would expect that the responses to these statements would be positively correlated; similarly, we would expect that the items (numbers 6 to 10) for the second attitude dimension would be intercorrelated. In fact, an implict assumption of factor analysis is that the correlations among items are *due* to their *common association* with the underlying factor. Alternatively, if the hypothesized grouping of the statements is appropriate, we would expect that the correlations between the statements *across* the scale groups would be relatively small and inconsequential.

The correlation matrix for this analysis is presented in Table 2.

TABLE 1

POTENTIAL ITEM POOL FOR CONSUMER ATTITUDE SCALES

Scale	Item No.	Statement	Mean	Standard Deviation
GOVERNMENT NON-INTERVENTION:				
	1.	The government should test competing brands of products and make the results of these tests available to consumers.	2.38	1.24
	2.	The government should set minimum standards of quality for all products sold to consumers.	2.58	1.21
	3.	In general, self regulation by business itself is preferable to stricter control of business by the government.*	3.25	1.10
	4.	The government should exercise more responsibility for regulating the advertising, sales, and marketing activities of manufacturers.	2.82	1.19
	5.	Government price control is the most effective way of keeping the prices of consumer products at reasonable levels.	3.29	1.19
CONSUMER ORIENTATION:				
	6.	Manufacturers do not deliberately design products which will wear out as quickly as possible.*	2.82	1.10
	7.	The exploitation of consumers by business firms deserves more attention than it receives.	2.19	.88
	8.	Despite what is frequently said, "Let the buyer beware" is the guiding philosophy of most manufacturers.	2.89	.97
	9.	Manufacturers often withhold important product improvements from the market in order to protect their own interests.	2.48	.94
	10.	Most manufacturers are more interested in making profits than in serving consumers.	2.11	.92

*These items were reversed scored.

TABLE 2
PRODUCT MOMENT CORRELATIONS FOR RESPONSES TO STATEMENTS

Correlations

Item Number:†	1	2	3	4	5	6	7	8	9	10
1	1.00									
2	.57	1.00								
3	.31	.32	1.00							
4	.51	.54	.46	1.00						
5	.50	.46	.34	.48	1.00					
6	.10	.05	.12	.06	.03	1.00				
7	.27	.27	.20	.28	.19	.29	1.00			
8	.13	.04	.10	.15	.09	.22	.25	1.00		
9	.10	.05	.07	.16	.06	.25	.25	.23	1.00	
10	.12	.12	.19	.26	.13	.29	.22	.30	.32	1.00

† These numbers correspond to the numbers and associated statements displayed in Table 1.

In general, the pattern of correlations shown in Table 2 is consistent with the logic discussed above. Note that the correlations for the items in the first group (marked off with the triangle in the upper left hand portion of the correlation matrix) are all positive and moderately strong; in the same vein, the correlations for the second grouping of statements (marked off with the triangle in the lower right hand corner of Table 2) are positive and significant, although they are somewhat lower than the correlations among the statements in the first grouping. On the other hand, consider the correlations between items from across the two groups (in the lower left hand rectangle of Table 2): most of the correlations are close to zero and therefore consistent with expectation. But, there is one exception; statement number 7 (which was expected to be part of the second group) is also positively correlated with the attitude items from the first group. This gives us a hint that this item may not be appropriate for the intended purposes. The factor analysis of the correlation matrix provides a direct and useful way of evaluating this concern.

Extracting Underlying Factors

There are different ways in which the factor analysis of the correlation matrix (Table 2) might be accomplished. Unfortunately, there are no generally accepted rules (or statistical criteria) for picking the optimal type of factoring method to be used in different situations and in fact there is considerable disagreement on this topic. In this study, however, the procedure used is principle factoring with interaction. Details on the computational aspects of this type of solution are provided elsewhere [12, 20, 22] so they will not be repeated here; but, it is worth noting that this approach is of the common factor analysis family of methods, as contrasted with the strict principle components methods. The name of the procedure is derived from the fact that it "iterates" toward a good solution by modifying the communality estimates; communality estimates will be discussed later in this article.

This factoring method used here was selected for two major reasons. First, it is a method which in the past has generally provided good initial factor solutions for scale development applications, and thus it is a procedure that is widely accepted and used. Also, most of the widely available computer statistical packages [6, 9, 20] offer this procedure as a computational option, and therefore it is one which readers will have access to for their own research.

Consideration of several summary indices or statistics which (1) characterize the initial factor solution and (2) suggest how many underlying factors will be studied in detail is a primary step toward interpretation. Table 3 presents several statistics relevant to these decisions. From Table 3 it may be noted that the initial factoring extracts 10 factors (i.e., as many factors as variables). However, factors are extracted in a mechanical fashion, and thus not all of them are necessarily important or meaningful.

One important and frequently used way of determining the importance of the successive factors is to evaluate the *eigenvalues* (sometimes called the eigenroot or characteristic root) associated with each factor. In general, each eigenvalue is a summary index of how much of the variance in the initial correlation matrix is accounted for by the associated factor. For example, the eigenvalue for the first factor in this analysis is 3.24, and accounts for 32.4% of the variability in the original correlation matrix. The relationship between these two numbers is direct. The percent of variance accounted for by a factor is the ratio of its eigenvalue to the total variance. With standardized variables, the variance of each variable is 1.0. Therefore, the total variance is 10.0 because there are ten statements. The ratio of the eigenvalue to the total, 3.24/10.0, gives the percent of variance for the factor.

The eigenvalue for the second factor is 1.7, and thus explains 17.1% of the variance. Because the factors are derived so that they are orthogonal (uncorrelated), the total (cumulative) variance explained by the first two factors is the sum of 32.4 and 17.1, or 49.5% of the variance.

Note from Table 3 that each of the successive factors explains a decreasing percent of the variance, and more specifically that the

TABLE 3
EIGENVALUES AND PERCENT OF VARIANCE EXPLAINED
FROM INITIAL FACTOR SOLUTION

Factor Extracted	Eigenvalue	Percent of Variance	Cumulative Percent of Variance
1	3.23733	32.4	32.4
2	1.70715	17.1	49.5
3	0.83316	8.3	57.8
4	0.79220	7.9	65.7
5	0.77344	7.7	73.4
6	0.67504	6.8	80.2
7	0.61671	6.2	86.4
8	0.53956	5.4	91.8
9	0.42519	4.2	96.0
10	0.40019	4.0	100.0

eigenvalues for factors 3 through 10 are all less than 1.0. Thus, only the first two factors explains more variance than any *one* of the initial variables. In determining how many factors to evaluate in detail researchers frequently concentrate on factors with eigenvalues greater than 1.0. For this reason, and because it is logically consistent with the a *priori* content groupings which motivated the analysis, the material which follows will focus on the structure and meaning of these two leading factors.

The Structure of the Leading Factors

Although the preceding evaluation suggests that there are in fact two major underlying dimensions or factors which summarize much of the variation among the consumer responses to the statements, this does not imply that the association among the statements and the two factor constructs is empirically consistent with the organization displayed in Table 1. Table 4, on the other hand, provides statistics which help the researcher to evaluate this issue and the nature of the factors.

The first columns in Table 4 display the factor structure ("loadings") matrix for the initial factor solution. Interpretation of the factor structure matrix, and subsequent understanding of its broader meaning, is most intuitive when the loadings are thought of as correlations between each of the initial variables and the resulting factor (composite) scores. It should be remembered that the factor scores are simply weighted combinations of all the original variables; that is, a factor score is a single (scalar) number itself. The variable "weights" (factor coefficients) for the linear combinations derived in this analysis are given in Table 5. Specifically, the factor score for an individual is computed by multiplying his or her standardized score for a variable times the associated factor coefficient, and then summing across all the variables; this is repeated for each factor.

Thus, considering the first row of the loadings matrix in Table 4, we see that the responses to statement 1 are correlated .69 and −.22 with the first and second factors, respectively. More generally, it may may be observed from Table 4 that the first five statements tend to correlate more substantially with the first factor than the second, and the second five (again with the exception of statement 7) correlate higher with the second factor than the first.

It should be noted that this initial factor pattern is intuitively consistent with the earlier discussion of the correlation matrix and a more general characteristic of factor analysis. Specifically, the highest correlations in the matrix (Table 2) were among the first five statements; thus, because the first factor extracted in a factor solution is

TABLE 4
FACTOR STRUCTURE (LOADINGS) MATRIX FOR INITIAL
AND VARIMAX ROTATED FACTOR SOLUTION

Item Number†	Initial Solution Loadings			Varimax Rotated Solution Loadings		
	Factor 1	Factor 2	Communality	Factor 1	Factor 2	Communality
1	.69	−.22	.52	.71	.11	.52
2	.67	−.29	.54	.73	.04	.54
3	.51	−.06	,26	.48	.17	.26
4	.75	−.15	.58	.73	.20	.58
5	.61	−.24	.43	.66	.06	.43
6	.25	.45	.27	.02	.52	.27
7	.46	.27	.28	.29	.44	.28
8	.28	.39	.23	.08	.47	.23
9	.27	.44	.27	.05	.51	.27
10	.38	.44	.34	.14	.57	.34

† These item numbers correspond to the numbers and associated statements displayed in Table 1.

TABLE 5
FACTOR SCORE COEFFICIENTS FOR FACTOR SOLUTION

Item†	Coefficients	
	Factor 1	Factor 2
1	.259	−.019
2	.291	−.087
3	.104	.031
4	.305	.045
5	.215	−.049
6	−.039	.260
7	.027	.202
8	−.025	.217
9	−.042	.250
10	−.026	.300

† Item numbers correspond to numbers and statements displayed in Table 1.

the one which explains the greatest portion of variation in the total matrix, it is not surprising that this factor is most highly characterized by these five statements. In addition, the correlations for other variables are relatively high on the first factor. This is not unexpected either; frequently the first factor extracted in the initial solution is substantially loaded on many (or all) of the original variables.

The communalities for each of the statements for the initial factor solution are also displayed in Table 4. In general, these indices suggest how much of the variance in the associated (original) variable is captured by the major factors. For example, 52% of the variance in the responses to statement 1 is explained by these two factors. The reader may confirm that each communality estimate is equal to the sum of the squared loadings for the associated variable. For example, the communality estimate for statement 2 is $(.67)^2 + (-.29)^2 = .54$. Moreover, the logic of this basic relationship is not complicated. The correlation of statement 2 with the first factor is .67. Thus the variance explained in item two by the first factor is $(.67)^2$; similarly, the variance explained by the second factor is $(-.29)^2$. Finally, because the factors are derived so that they are *uncorrelated*, the *total* variance explained in the item is simply the *sum* of the variance explained by each of the factors individually.

Further examination of the initial factor structure matrix reveals several patterns which warrant additional attention. First, most of the items have associated loadings which are quite large on more than one factor. And, as a related issue, several of the items (see for example items 8 and 10) have loadings which are about equal in size. Such "split" loadings make it difficult to clearly interpret which variables "belong" on a factor, and thus which can be legitimately grouped together. This is a common occurence in initial factor analysis solutions, and it has been found that frequently such initial factor solutions can be simplified or "upgraded" by rotation of the loadings matrix.

Rotation

Just as there are many different ways of extracting the initial factor solution, there are a variety of useful approaches to factor rotation. The choice of a rotational procedure is generally determined by the researcher's objectives. In this case, a Varimax rotation procedure [16] was chosen because it is one which generally substantially simplifies the loadings structure and makes the meaning of the factors most clear. To accomplish this, the Varimax rotation seeks to "reorganize" the loadings on the factors so that a factor is characterized only by variables with high loadings (approaching 1.0) or low loadings (approaching 0.0).

The rotated factor structure matrix for this analysis is displayed in Table 4 along with the unrotated solution to facilitate comparison and contrast. Note that the rotated loadings structure has in fact been "simplified;" most variables now load (correlate) substantially on only one factor. For example, in the initial (unrotated) solution statement 10 has loadings of .38 and .44; it was substantially correlated with both factors. However, in the *rotated* solution the loading on the first factor is smaller, and on the second factor it is larger (.14 and .57 respectively). By contrasting the rotated and unrotated loadings in Table 4, it can be seen that the factorial simplicity of each of the variables, except 3 and 4, has been improved. On these two variables that were not improved, the rotated loadings still load substantially higher on the first function than on the second function. In general, then, the Varimax rotation accomplished the objective of simplifying the factor structure.

Before moving to discussion and interpretation of the factors, it should be noted that the rotation process, although simplifying the pattern of loadings, does not change the *total* amount of variance explained by the factors; it simply redistributes it across the factors. For example, note that the communalities in Table 4 for the rotated and unrotated solutions are the same; the individual loadings have been changed, but the relationship *among* the *variables* with respect to the factors has not changed and the *sum* of the squared loadings for a specific variable is still the same. This point is substantiated by visual presentation in Figure 1 and Figure 2. In each figure, the axes represent the factors, and the points in the factor "space" correspond to the original statements from Table 1. The coordinates used to position the statements in the factor space are simply the loadings from the initial (Figure 1) or rotated (Figure 2) factor matrices. By examining the figures, it may be seen that the positions of the variables *relative to each other* in the factor space has not been changed by the rotation. Similarly, as depicted by the dotted lines in Figure 1, only the axes (factors) have been rotated. As a general note, plotting variables and factors in this way is also frequently helpful in interpreting the factors, and in some cases in deciding what type of rotation to use to simplify the structure of the loadings.

The Factors

Now that the simplified factor structure matrix is available, it is possible to characterize the underlying factors and to determine which statements should be included with each factor. The loadings for the statements on the first factor are consistent with the expectation reflected by the grouping of statements in Table 1. Each of these statements relates to a different aspect of government intervention in the market place, and the factor structure appears consistent with the supposition that an attitude toward government intervention in business practices explains the correlations among the statements. Because of the way in which the variables are scored, a low score on

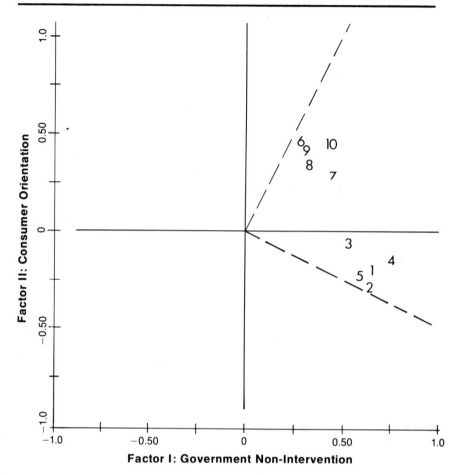

Figure 1—Unrotated Solution

this factor reflects a favorable attitude toward government intervention, and a high score suggests an underlying attitude which is
unfavorable to intervention. Thus, we label this a government non-
intervention factor.

The loadings associated with the second grouping of statements
are all higher on the second factor than on the first factor. However,
in general the largest loadings for the second factor are not as high
as the largest loadings on the first factor. Furthermore, statement 7
is substantially correlated with both the first factor (.29) *and* the
second factor (.44). Thus, this variable is somewhat problematic; even
after rotation, it still has split loadings across the two factors. This is,
of course, apparent in the graphic representations (Figure 1 and

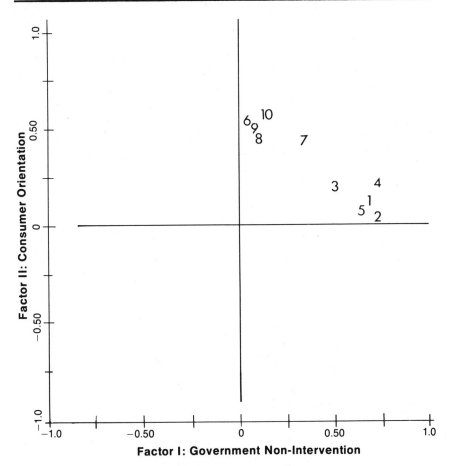

Factor I: Government Non-Intervention

Figure 2—Varimax Solution

Figure 2) where this variable is visibly separated from the cluster of statements which are closely aligned with the "consumer-orientation" of business factor (vertical axis).

It may be remembered (Table 1) that statement 7 suggests that "the exploitation of consumers by business firms deserves more attention than it receives." Analysis of the content of this statement suggests one probable reason why it is loaded on both factors. While the statement does in fact reflect aspects of respondents' attitudes concerning the consumer orientation of the firm (factor 2), it may simultaneously be that respondents do not think that the attention should be in the form of government regulation (factor 1).

Therefore, the factor analysis results further confirm the suspicion raised by the pattern of raw correlations previously discussed; although statement 7 is related to other attitudes concerning the consumer orientation of businesses, it is confounded in that it is related simultaneously to attitudes concerning government intervention. Because it is our intention to develop scales which reflect these two underlying attitudes in a way that is not confounded, this statement should be excluded in developing the composite measure. Worded another way, statement 7 would decrease the validity of a composite measure developed from these statements because it is also related to an underlying attitude it is *not* intended to measure.

In this study, attitude scales were developed by averaging responses to individual statements. For example, a consumer's attitude concerning government intervention was represented by his or her average response to the first five items in Table 1; in the same vein, the consumer orientation of business attitude was measured by the average of responses to statements 6, 8, 9, and 10. It should be noted that in some research applications the actual factor scores are used as the composite scale measures; the procedure used here is chosen primarily because it is consistent with tradition in attitude research.

ATTITUDES OVER TIME

Now that the individual statements and their interrelationships have been evaluated, and summary attitude measures have been developed, it is possible to evaluate the trend in these underlying attitudes over time. After the two attitude scores were computed for each respondent in each of the four samples over the seven year period, the mean score for men and the mean score for women on each of the measures was computed for each survey year. Figure 3 provides a visual summary of these statistics. In addition, a repeated measures analysis of variance was computed for the data; year and sex of respondent were treated as factors, and the newly constructed attitude measures were treated as the repeated measures. The discussion that follows is based on the repeated measures analysis, but the reader may confirm the interpretations offered by inspection of Figure 3.

The most notable pattern in the results is the significant difference over time in the attitude of the respondents to government intervention. In recent years, both men and women report an increased disposition for non-intervention on the part of the government. Further, there is a significant difference between the responses of men and women: across time, men are less favorable to intervention than

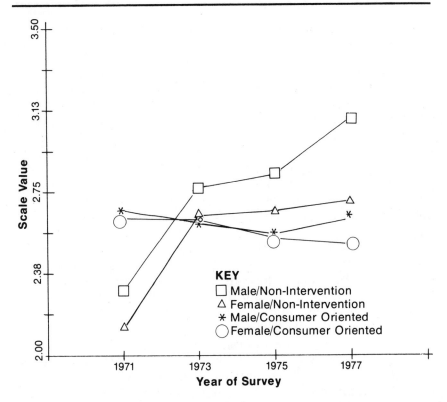

Figure 3—Consumer Attitudes Over Time

are women. In addition, there is a sex by time interaction in the pattern of responses; as may be observed in Figure 3, differences in attitude on this dimension between men and women have become more pronounced over time.

By contrast, attitudes concerning the consumer orientation of business were relatively more stable; but, there were still significant shifts over time. In general, consumers in the 1977 survey have a less favorable attitude concerning the consumer orientation of firms than did their counterparts in earlier surveys. Here again, there are also significant response differences between men and women . . . especially in the 1977 survey. Men are generally more favorable in their view of business than are women.

In summary, then, it appears that consumers now have less faith in government intervention and controls than they did during the early years of the consumer movement, and they still do not appear convinced that business is operating in line with the marketing concept and the motivation philosophy of a consumer orientation. It seems that consumers do not perceive that the changes during this period have helped them.

DISCUSSION

The factor analysis results presented in this report were based on the respondents to the 1977 survey. It should be noted, however, that the basic factor results were similar when the analysis was done on the data across all four surveys, as well as when the data from the other surveys were analyzed individually. In fact, in this research, a procedure known as Procrustes Rotation [23] was used to test the similarity of the factor results from each wave of the survey. Procrustes analysis is mentioned here because it can be a useful approach in evaluating the equality of factor solutions across different samples, but has not yet been widely used in business research.

In the same vein, in this paper we have limited discussion to the results of a Varimax rotation. It should be mentioned, however, that in research where the factor structure matrix is more complicated (e.g., when there are many variables being analyzed), it is sometimes difficult to chose the appropriate type of rotation. Mathematical psychologists have recently begun to work on this problem, and have suggested several indices which can help guide the researcher in picking the best rotation [7, 17]. For example, Kaiser [17] has proposed an index of factorial simplicity which can be used to compare the quality of factor structure matrices procedured by different rotations. In fact, Darden and Perreault [8] have shown that Kaiser's index can be advantageously applied when Q-type factor analysis procedures are used to develop taxonomies of consumers.

Earlier in this paper it was suggested (1) that scales properly constructed from multiple items were desirable because of improved properties of reliability and validity, and (2) that factor analysis was a useful input in constructing such scales. While it is beyond the scope of this paper to develop in detail the interrelationship of factor analysis and these scale properties, it should be noted that the mathematical foundations of this relationship have been clearly set forth. The interested reader will find both the appropriate derivations (and applied examples) in the work of Heise and Bohrnstedt [13].

In conclusion, this paper provides an illustration of the application and interpretation of factor analysis in business research. While it has been emphasised that there are many aspects of factor analysis that require the researcher to make judgments based on non-statistical criteria, its use should not be avoided on this basis. Instead, the researcher should realize both its potential benefits and limitations, and use it with those issues clearly in mind. When thoughtfully applied, it is a powerful and useful procedure that can help the researcher to better understand the complex interrelationships of attitudinal and behavioral data.

REFERENCES

1. Aaker, David A. and George S. Day. "Corporate Responses to Consumerism Pressures," *Harvard Business Review*, Vol. 50, No. 6 (November-December 1972), pp. 114–124.
2. Armstrong, J. Scott. "Derivation of Theory by Means of Factor Analysis or Tom Swift and His Electric Factor Analysis Machine," *American Statistician*, Vol. 21 (December 1967), pp. 17–21.
3. Barksdale, Hiram C. and William R. Darden. "Consumer Attitudes Toward Marketing and Consumerism," *Journal of Marketing*, Vol. 36, No. 4 (October 1972), pp. 28–35.
4. Barksdale, Hiram C., William R. Darden, and William D. Perreault, Jr. "Changes in Consumer Attitudes Toward Marketing, Consumerism and Government Regulation: 1971–1975," *Journal of Consumer Affairs*, Vol. 10, No. 2 (Winter 1976), pp. 117–139.
5. Barksdale, Hiram C. and William D. Perreault, Jr. "The Role and Impact of Corporate Consumer Affairs Professionals in Marketing Management," *Journal of Marketing*, Vol. 42 (1978), in press.
6. Barr, Anthony J., James H. Goodnight, John P. Sall, and Jane T. Helwig. *A User's Guide to SAS 76*. (Raleigh: The SAS Institute, 1976).
7. Bentler, Peter M. "Factor Simplicity Index and Transformations," *Psychometrika*, Vol. 42, No. 2 (June 1977), pp. 277–296.
8. Darden, William R. and William D. Perreault, Jr. "Classification for Market Segmentation: An Improved Linear Model for Solving Problems of Arbitrary Origin," *Management Science*, Vol. 24, No. 3 (November 1977), in press.
9. Dixon, W. J. (ed.). *BMD: Biomedical Computer Programs*. (Berkeley: University of California Press, 1973).
10. Farley, John U. "Why Does 'Brand Loyalty' Vary Over Products?" *Journal of Marketing Research*, Vol. 1, No. 4 (November 1963), pp. 9–14.
11. Greyser, Stephen A. and Steven L. Diamond. "Business Is Adapting to Consumerism," *Harvard Business Review*, Vol. 52, No. 5 (September-October 1974), pp. 38–58.
12. Harman, Harry H. *Modern Factor Analysis*. (Chicago: The University of Chicago Press, 1976).
13. Heise, David R. and George W. Bohrnstedt. "Validity, Invalidity, and Reliability," in E. F. Borgatta and G. W. Bohrnstedt (eds.), *Sociological Methodology*. (San Francisco: Jossey-Bass, Inc., 1970), pp. 104–129.
14. Herrman, Robert O. "Consumerism: Its Goals, Organization and Future," *Journal of Marketing*, Vol. 34, No. 4 (October 1970), pp. 55–60.
15. Johnson, Richard M. "Q Analysis of Large Samples," *Journal of Marketing Research*, Vol. 7, No. 1 (February 1970), pp. 104–105.
16. Kaiser, Henry F. "The Varimax Criterion for Analytic Rotation in Factor Analysis," *Psychometrika*, Vol. 23, No. 3 (September 1958), pp. 31–36.
17. Kaiser, Henry F. "An Index of Factorial Simplicity," *Psychometrika*, Vol. 39, No. 1, (March 1974), pp. 31–36.

18. Kotler, Philip. "What Consumerism Means for Marketers," *Harvard Business Review,* Vol. 51, No. 3 (May-June 1973), pp. 48–57.
19. Miles, Robert H. and William D. Perreault, Jr. "Organizational Role Conflict: Its Antecedants and Consequences," *Organizational Behavior and Human Performances,* Vol. 17, No. 1 (October 1976), pp. 19–44.
20. Nie, Norman H., C. Hadlai Hull, Jean G. Jenkins, Karin Steinbrenner, and Dale H. Bent. *Statistical Package for the Social Sciences.* (New York: McGraw-Hill, 1975).
21. Peter, J. Paul. "Reliability, Generalizability, and Consumer Behavior," in W. D. Perreault, (ed.), *Advances in Consumer Research, Volume IV.* (Atlanta: Association for Consumer Research, 1977), pp. 394–400.
22. Rummel, R. J. *Applied Factor Analysis.* (Evanston: Northwestern University Press, 1970).
23. Schönemann, Peter H. "A Generalized Solution of the Othogonal Procrustes Problem," *Psychometrika,* Vol. 31, No. 1 (1966), pp. 1–10.
24. Wells, William D. "Psychographics: A Critical Review," *Journal of Marketing Research,* Vol. 12, No. 2 (May 1975), pp. 196–211.
25. Wells, William D. and Jagdish N. Sheth. "Factor Analysis in Marketing Research," in R. Ferber (ed.), *Handbook of Marketing Research.* (New York: McGraw-Hill, 1971).

THE SUBSTITUTION CONTROVERSY: ATTITUDES OF PHARMACISTS TOWARD REPEAL OF ANTISUBSTITUTION LAWS

THOMAS R. SHARPE and MICKEY C. SMITH

The objectives of this investigation were to provide a parsimonious description of the attitudes of pharmacists toward drug antisubstitution laws and to identify those issues influencing the pharmacist's response in favor of or in opposition to repeal of antisubstitution laws.

To meet these objectives, a questionnaire was developed for

Thomas R. Sharpe, Ph.D., is Assistant Director for Administrative Sciences Research, Research Institute of Pharmaceutical Sciences; and Mickey C. Smith, Ph.D., is Professor and Chairman, Department of Health Care Administration, School of Pharmacy, The University of Mississippi, University, Mississippi 38677.

Revised from a paper presented before the Section on Economics and Administrative Science, Academy of Pharmaceutical Sciences, American Pharmaceutical Association, San Francisco, California, April 1975.

The authors gratefully acknowledge the financial support of the American Foundation for Pharmaceutical Education and the Research Institute of Pharmaceutical Sciences, University of Mississippi.
Reprinted by permission of the authors and publisher Drugs in Health Care *Vol. 3 (Winter 1976) pp. 21–34.*

data collection which included an Attitude Toward Antisubstitution Laws Scale (A Scale). The A Scale items were located in literature pertaining to the antisubstitution controversy. The survey instrument was mailed to a systematic random sample of 5,000 U.S. pharmacists. Usable returns were received from 847 (16.94%) respondents.

In order to provide a parsimonious description of attitudes of pharmacists toward antisubstitution laws, a principal components multiple factor analysis (varimax rotation) was performed on the A Scale items. Analysis revealed an eight factor solution which accounted for 60.39% of total variance.

To meet the second objective, varimax factor scores derived from the A Scale analysis were used as independent variables in a multiple discriminant analysis. The dependent variable was the respondent's categorical yes-no response to an item asking specifically whether he favored repeal of the laws. The discriminant function developed from the analysis sample yielded a statistically significant ($\alpha = 0.05$) hit-ratio of 87.74%. From this analysis it was concluded that, from the pharmacist's point of view, pharmacist autonomy appears to be the overriding issue in the antisubstitution controversy.

Since the American Pharmaceutical Association (APhA) officially committed itself in 1970 to seek repeal of drug antisubstitution laws, much attention has been given to the arguments favoring and opposing such repeal. Debate has focused on the controversial issues of therapeutic equivalency among branded and generic products; pharmacists' and physicians' professional prerogatives, abilities and duties; pharmacists' legal and professional liabilities; the role of the professional organizations (especially APhA) in the controversy; and the economics of generic name versus brand name dispensing.

Most research dealing directly with antisubstitution laws has focused on attitudinal measurements. Several studies have dealt only with pharmacists' attitudes,[1-4] while others have included physician and consumer attitudes.[5,6]

It is ironic that in taking opposite stands on the issue, both APhA and the National Association of Chain Drug Stores (NACDS) may not have represented the majority viewpoint of their constituents. McCormick found that a majority of pharmacists practicing in chain pharmacies *favored* repeal, while a majority of pharmacists practicing in independent community pharmacies *opposed* repeal.[5] Similarly, while hospital pharmacists voted as a block in favor of repeal at the 1970 APhA Convention, only 34% of hospital pharmacists in this study favored repeal of the laws.[5]

With the exception of the studies by Nelson[6] and McCormick,[5] the attitudinal research to date has consisted primarily of "head counts" of those who favor or oppose repeal of the laws. Little has been provided in the way of describing the dimensionality of these attitudes. Further, while the pharmacy and medical literature has been replete with arguments both for and against repeal, in delineating those arguments it has provided scant insight into which arguments are relevant to (*i.e.*, predictive of) the individual's attitude in favor of or in opposition to repeal of antisubstitution laws.

The objectives of this investigation, then, were as follows:

1. To provide a parsimonious description of attitudes of pharmacists toward antisubstitution laws.
2. To identify those issues which predict the pharmacist's response favoring or opposing repeal of antisubstitution laws.

METHODOLOGY

In order to meet the study objectives, a questionnaire was developed for data collection. This instrument consisted of three sections: a Professionalism Scale (P Scale), an Attitude Toward Antisubstitution Laws Scale (A Scale) and Structure of Practice Data (demographics, description of practice setting, etc.). The A Scale and its development are discussed below.

The A Scale items were selected from pharmaceutical and medical literature relevant to the antisubstitution controversy. Located from these sources were 205 varied statements both supporting and opposing antisubstitution laws. The statements were then placed into eight categories: six pertained to theoretical professional dimensions; two additional categories, created to classify the remaining statements, pertained to economic and therapeutic dimensions. Next, 39 items were selected to represent the "universe" of the 205 items in eight categories. Bases for this selection were the investigator's judgments as to an item's relevance to the research problem, and the item's representativeness of the category into which it had been placed. An item asking specifically whether the respondents favored repeal of antisubstitution laws was added to the 39 items. Item sequence was randomly assigned to preclude order bias; the format represented a seven-point, agree-disagree, Likert-type scale.

The instrument was pretested by administering it to 38 community and hospital pharmacy practitioners who represented both rural and metropolitan practice settings. Each pharmacist was asked to complete the questionnaire according to the written instructions; in addition, he/she was to place an "X" adjacent to any instruction or item which seemed unclear. After the pharmacist completed the questionnaire, the investigator conducted a probing interview to determine why a given instruction or item was ambiguous. As a result, several items were modified to enhance their clarity.

After this pretest and subsequent minor modifications, the questionnaires with cover letters, explaining the study's purpose and requesting the pharmacist's participation, and postage-paid reply envelopes were sent to a systematic random sample of 5,000 United States pharmacists. The pharmacist population consisted of all licensed pharmacists (approximately 100,000 in 50 states) on the revised list of the National Association of Boards of Pharmacy (NABP). Posted from Chicago, Illinois on July 16, 1974, the questionnaires were mailed by a medical mailing service which maintains the list of pharmacists for NABP. Usable returns were received from 847 (16.94%) respondents.

The data were keypunched on cards suitable for computer analysis and screened for missing values. Missing values were identified in the A Scale in 43 (0.13%) instances. The distribution of missing values appeared to be random among the items. A value was randomly assigned to each item which had been left blank by the respondent. The data were then analyzed according to the procedure illustrated in Figure 1.

ANALYSIS OF A SCALE

To provide a parsimonious description of attitudes of pharmacists toward antisubstitution laws, a principal components multiple factor analysis, using the varimax criterion for factor rotation, was performed on the items in the A Scale. The initial analysis of the 39 items incorporated an eigenvalue-one criterion for determining the number of factors to extract.

Factor analysis was chosen as the appropriate technique because of its expressed purpose of data reduction and summarization. The method allows the researcher to analyze the interrelationships among variables (e.g., scale items) in terms of their common underlying dimensions (factors). These factors may be considered the essential determining constructs representing a new set of variables which are defined solely in terms of the original dimensions.

If the researcher's goal is to develop predictive equations, principal components analysis is the appropriate technique.[8] This is based on the characteristic that principal components analysis carries the total variance into the factor matrix for subsequent extraction of the appropriate factor structure.

An orthogonal factor rotation was chosen because it results in both factor loadings and factor scores which are independent. Thus, if these factor scores were to be used as predictor variables in a subsequent analysis, they would meet the assumption that predictor variables are uncorrelated.

The varimax criterion for orthogonal rotation was selected because it maximizes the number of very high and very low factor loadings, thus providing the simplest factor structure solution. As such, varimax is generally accepted as the best analytic orthogonal

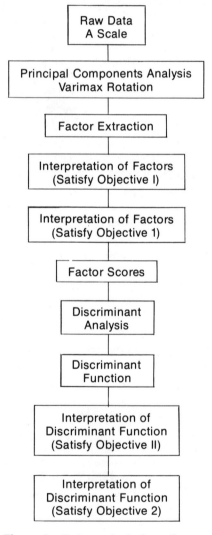

Figure 1—Data analysis flow diagram

rotation technique.[9-11] The computer program FACTOR of the Veld-man Statistical Package [12] was used for the analysis.

Based on initial analysis, two items were deleted because of their low (less than 0.40) communalities. A third item was deleted because it represented a single-item factor and, therefore, was inconsistent with the objective of parsimony.

Utilizing the eigenvalue-one criterion, the principal components analysis (varimax rotation) of the remaining 36-items resulted in an eight factor solution which accounted for 60.39% of the total variance. The solution provided one of the most clearly interpretable factor structures to that point. An additional factor analysis constraining the solution to nine factors resulted in a factor structure which provided no advantage over the eight factor solution, and a seven factor solution was distinctly less interpretable. Consequently, the eight factor solution of the 36-item A Scale was selected as the best solution.

Column sum of squares (eigenvalue) and percent of trace extracted with respect to each factor in the unrotated principal components solution are presented in Table 1. Additionally, Table 2 indicates the rotated factor matrix derived from this solution. The matrix includes the factor loadings and communality for each item in the 36-item A Scale. To aid the reader, the highest factor loading for each item has been underlined.

Factors were labeled according to factor loading patterns of the scale items. The highest loadings were considered most important for the purpose of deriving labels. Thus, the factors were named as follows:

Factor I Competence for Brand Selection
Factor II Physician Autonomy/Derivation of Pharmacist Prerogatives

TABLE 1
A SCALE EIGENVALUES AND PERCENT OF TRACE EXTRACTED FOR
UNROTATED PRINCIPAL COMPONENTS FACTOR SOLUTION

	Factors								
	I	II	III	IV	V	VI	VII	VIII	Total
Column Sum of Squares (Eignevalue)	10.46	2.26	2.07	1.87	1.46	1.45	1.11	1.06	
Percent of Trace	29.05	6.27	5.76	5.21	4.05	4.03	3.08	2.95	60.39

TABLE 2
A SCALE FACTOR LOADINGS FOR ROTATED PRINCIPAL COMPONENTS FACTOR MATRIX

Item No.*	Factors								Com-munality
	I	II	III	IV Trace = 36.00	V	VI	VII	VIII	
1	0.0545	-0.1402	-0.1223	0.7228	-0.1085	0.0491	0.2029	0.0647	0.6195
2	-0.0343	0.8145	0.0234	-0.0971	0.0016	-0.0345	-0.1793	-0.0833	0.7148
3	0.0518	-0.0452	-0.7696	0.0980	-0.1383	0.0791	0.1211	0.0211	0.6470
4	0.0916	0.0564	-0.1869	0.3716	-0.5621	0.0731	0.3279	0.1086	0.6252
5	0.0107	0.0608	-0.0635	0.1273	-0.2435	0.6297	0.1486	0.0016	0.5019
6	-0.0103	0.8268	0.0435	-0.0883	-0.0316	-0.0331	-0.1667	-0.0955	0.7324
8	0.4036	-0.0424	-0.1302	-0.0099	-0.1199	0.4791	0.0022	-0.0537	0.4285
9	0.1576	-0.2457	-0.2346	0.1591	-0.2584	0.1074	0.5596	0.1301	0.5740
12	0.5658	0.0121	-0.1451	0.1779	-0.1670	0.1083	0.3154	0.0216	0.5126
13	0.1249	0.0300	-0.2156	0.4820	-0.4305	0.1141	0.3949	0.1068	0.6609
14	0.2714	0.0508	-0.2164	0.1233	-0.5784	0.3014	0.1585	-0.0854	0.5961
15	-0.1645	0.1025	0.1240	0.0532	0.7163	-0.1112	-0.1750	-0.0352	0.6131
16	0.3151	-0.0597	-0.0962	-0.0404	-0.1157	0.1707	0.5127	0.1861	0.4538
17	0.1097	-0.0405	-0.0227	0.0028	-0.0374	0.7906	0.0908	0.0789	0.6551
18	0.1390	-0.1458	-0.1074	0.8155	0.0229	0.1022	0.1164	-0.0162	0.7419
19	-0.0064	0.1544	0.7441	-0.1187	0.2180	-0.0095	-0.1453	-0.1055	0.6716
20	-0.0460	0.2296	0.0024	-0.1037	0.2099	0.0280	-0.0223	-0.7630	0.6931
21	0.2934	-0.1930	-0.1212	0.1096	-0.1149	0.1624	0.6776	0.0004	0.6487

Item	F1	F2	F3	F4	F5	F6	F7	F8	h²
22	0.1729	-0.1302	-0.1406	0.1540	-0.1198	0.1199	0.7328	-0.0223	0.6566
23	-0.6044	0.2517	0.0546	-0.1294	0.0430	0.0842	-0.2580	-0.1399	0.5435
24	-0.2036	0.5857	0.1849	-0.0548	0.1586	0.0156	-0.2873	-0.0921	0.5381
25	0.1187	-0.0395	-0.8166	0.1307	-0.1085	0.0403	0.1429	0.0211	0.7338
26	0.2511	-0.1104	-0.1919	0.1795	-0.1273	0.1308	0.7139	0.0289	0.6882
27	0.6510	-0.1831	-0.0642	0.0924	-0.0964	0.0223	0.4134	-0.0521	0.6535
28	0.1843	-0.0869	-0.1441	0.8105	-0.0088	0.0587	0.1525	-0.0153	0.7462
30	-0.1047	0.4912	0.0565	-0.0289	0.3869	0.0490	0.1181	0.0806	0.4288
31	0.7170	-0.0509	-0.0051	0.0612	-0.0965	0.1355	0.1272	0.1232	0.5797
32	0.2776	0.0205	-0.1398	0.0348	0.0748	0.1449	0.2564	0.6737	0.6445
33	0.5803	0.0517	-0.0982	0.2535	-0.1206	0.1230	0.2153	0.1080	0.5010
34	0.3493	0.0393	-0.1231	0.3026	-0.1296	0.1355	04351	0.1768	0.4860
35	-0.0819	0.1744	0.1804	-0.0265	0.5982	-0.1804	-0.1510	-0.0217	0.4841
36	-0.1477	-0.0043	0.0208	-0.3825	0.3078	0.2530	0.0314	-0.2126	0.4135
37	0.2881	-0.0971	-0.2259	0.2560	-0.2458	0.0949	0.6493	0.0695	0.7049
38	0.1311	0.0246	-0.7446	0.1275	-0.0906	0.1241	0.1970	0.0038	0.6508
39	0.0631	-0.0459	-0.1456	0.0576	-0.0614	0.7906	0.1795	0.0578	0.6950
40	0.0258	0.1185	0.0757	-0.1126	0.5223	-0.0232	-0.4531	-0.1762	0.5428
Percent of Variance (Varimax Rotation)	8.40	6.60	8.16	8.00	7.62	6.50	11.38	3.73	60.39

*The number of each item indicates its location in the A Scale. (Note that items 7, 10, 11 and 29 are not included in this table, since they were excluded from analysis. See text for explanation).

Factor III   Therapeutic Equivalency
Factor IV   Economics of Brand Selection Control
Factor V   Ethics of Brand Selection
Factor VI   Identification and Professional Regulation of Brand
             Selection Abusers
Factor VII   Pharmacist Autonomy/Essential Pharmacist Services
Factor VIII Role of the Professional Organization

These factors, including their item loadings, are presented in Tables 3–10.

Stratified reliability coefficients were then obtained based on the method outlined by Kerlinger.[13] This method uses two-way (items vs. individuals) analysis of variance in the calculation procedure. These are shown in Table 11. The table indicates that the reliability coefficients ranged from 0.4358 to 0.8776, with an overall coefficient of 0.8855.

DEVELOPMENT OF DISCRIMINANT FUNCTION

To meet the second objective of the study, a discriminant function was developed which would distinguish those respondents who supported antisubstitution laws from those who opposed them. From the usable returns, 610 (72.02%) of the pharmacists supported repeal, while 237 (27.98%) opposed repeal of the laws. Varimax factor scores * for each respondent on each of the eight factors derived from the A Scale were input as independent variables in the multiple discriminant analysis (MDA). The dependent measure was the respondent's categorical yes-no response to a question which asked specifically whether the pharmacist favored repeal of the laws. The computer program DSCRIM of the Veldman Statistical Package [12] was used for the analysis.

Before submitting data to MDA, however, respondents were randomly assigned to two groups in a ratio of 3:1. The analysis sample, consisting of 635 respondents, was used to develop the discriminant function. The smaller group of 212 respondents, representing the "hold-out" sample, was used to construct a classification matrix to

---

* The varimax factor scores derived by the Veldman computer program represented exact factor scores obtained directly from the varimax rotation of the components model according to the formula: $Y_{NK} = Z_{NM} V_{MK} I_{\Delta K} K^{-2} V_{KM} W_{MK}$, where V is a matrix of unrotated loadings, E is a diagonal matrix of roots and W is the varimax-rotated matrix of loadings. This formula employs the matrix of loadings and the matrix of raw scores standardized by columns (variables). As such, the factor scores in this case are exact, orthogonal and standardized ($X = 0$, $\sigma = 1$).

TABLE 3
FACTOR I: COMPETENCE FOR BRAND SELECTION

Item No.[a]	Loading	Item[b]
12	0.5658	The pharmacist has the education and training to make the brand selection decision.
23	-0.6044	The physician is more competent than the pharmacist to make the brand selection decision.
27	0.6510	The pharmacist is more competent than the physician to make the brand selection decision.
31	0.7170	Pharmacists are more aware of problems with bioavailability and therapeutic equivalence than most physicians.
33	0.5803	Pharmacists are more aware of drug recalls than physicians and in that respect are in a better position to choose the manufacturer who is most likely to produce a quality product.

[a]The number of each item indicates its location in the A Scale.
[b]Items included were those which had their highest loadings on Factor I.

TABLE 4
A SCALE FACTOR II: PHYSICIAN AUTONOMY/DERIVATION OF
PHARMACISTS' PROFESSIONAL PREROGATIVES

Item No.[a]	Loading	Item[b]
2	0.8145	Pharmacists should not substitute one drug product (brand) for another if it irritates the physician.
6	0.8628	Pharmacists should not substitute one drug product (brand) for another if physicians are opposed to it.
24	0.5857	It is the physician's professional right to choose the drug product (brand).
30	0.4912	The pharmacist must earn the respect of the physician before he can substitute one drug product (brand) for another.

[a]The number of each item refers to its location in the A Scale.
[b]Items included were those which had their highest loadings on Factor II.

TABLE 5
A SCALE FACTOR III: THERAPEUTIC EQUIVALENCY

Item No.[a]	Loading	Item[b]
3	-0.7696	Lack of therapeutic equivalency among drug products (brands) has been exaggerated.
19	0.7441	The extent of significant therapeutic differences among chemically equivalent drug products is great.
25	-0.8166	There are relatively few instances of therapeutic non-equivalence among different brands of the same drug product.
38	-0.7446	In most cases, drug products which meet official standards are therapeutically equivalent.

[a]The number of each item refers to its location in the A Scale.
[b]Items included were those which had their highest loadings on Factor III.

TABLE 6
A SCALE FACTOR IV: ECONOMICS OF BRAND SELECTION CONTROL

Item No.[a]	Loading	Item[b]
1	0.7228	Repeal of antisubstitution laws would result in lower operating costs for the pharmacist.
13	0.4820	Repeal of antisubstitution laws would lower prescription costs to patients.
18	0.8155	Repeal of antisubstitution laws would allow the pharmacist to reduce the number of products in his inventory, thus reducing his costs.
28	0.8105	Repeal of antisubstitution laws would result in decreased cost to the pharmacist through quantity purchases of fewer brands.
36	-0.3826	Prescription costs to the patient would increase if antisubstitution laws were repealed.

[a]The number of each item refers to its location in the A Scale.
[b]Items included were those which had their highest loadings on Factor IV.

TABLE 7
A SCALE FACTOR V: ETHICS OF BRAND SELECTION

Item No.[a]	Loading	Item[b]
4	-0.5621	If antisubstitution laws were repealed pharmacists would pass at least part of their savings on to the patient.
14	-0.5784	Most pharmacists are dedicated professionals and would not abuse the right to substitute one drug product (brand) for another.
15	0.7163	Repeal of antisubstitution laws would result in increased profits to the pharmacist through dispensing of cheap, possibly inferior, products.
35	0.5982	There are many business-oriented pharmacists who would misuse the right to substitute one drug product (brand) for another.
40	0.5223	Pharmacists who favor repeal of antistitution laws do so for purely personal economic gains.

[a]The number of each item refers to its location in the A Scale.
[b]Items included were those which had their highest loadings on Factor V.

TABLE 8
A SCALE FACTOR VI: IDENTIFICATION AND PROFESSIONAL REGULATION OF BRAND SELECTION ABUSERS

Item No.[a]	Loading	Item[b]
5	0.6297	The pharmacy profession itself would penalize those who substitute clearly inferior products if brand substitution were allowed.
8	0.4791	The products which vary widely in therapeutic effect are well known to the pharmacist.
17	0.7906	If an individual pharmacist were to substitute one drug product (brand) for another in a manner which is incompetent or inconsistent with the patient's interest, my colleagues could easily identify him.

[a]The number of each item refers to its location in the A Scale.
[b]Items included were those which had their highest loadings on Factor VI.

TABLE 9
A SCALE FACTOR VII: PHARMACIST AUTONOMY/ESSENTIAL
PHARMACIST SERVICES

Item No.[a]	Loading	Item[b]
9	0.5596	To deny the pharmacist this option (to substitute) is a detriment to the patient.
16	0.5127	Pharmacists should not be intimidated into relinquishing their inherent responsibilities for drug product (brand) selection.
21	0.6776	It is the pharmacist's professional right to choose the drug product (brand).
22	0.7328	Antisubstitution laws prevent the pharmacist from providing his full complement of services to the patient.
26	0.7139	Antisubstitution laws limit the prerogatives of the pharmacist in exercising his professional role.
34	0.4351	Pharmacists can offer an essential service to their patients by dispensing the least expensive but therapeutically effective drug product (brand).
37	0.6493	It is in the public interest for the pharmacist to have the option to substitute.

[a]The number of each item refers to its location in the A Scale.
[b]Items included were those which had their highest loadings on Factor VII.

TABLE 10
A SCALE FACTOR VIII: ROLE OF THE PROFESSIONAL ORGANIZATION

Item No.[a]	Loading	Item[b]
20	-0.7630	The APhA shouldn't take a stand on antisubstitution since it doesn't represent a majority of pharmacists anyway.
32	0.6737	My professional association should take a leadership position on the antisubstitution issue.

[a]The number of each item indicates its position in the A Scale.
[b]Items included were those which had their highest loadings on Factor VIII.

test the predictive validity of the discriminant function. If all observations had been used to calculate the discriminant function and then classified with the function, there would have been an upward bias in the number of correctly classified entities.

The discriminant function derived from the analysis sample took the following form:

$$Z = 0.2387X_1 - 0.1608X_2 - 0.2667X_3 + 0.3107X_4 - 0.2670X_5$$
$$+ 0.1266X_6 + 0.7939X_7 + 0.2787X_8$$

Where: Z = the discriminant score (Z-score)

X_1 = factor score on Factor I
X_2 = factor score on Factor II
X_3 = factor score on Factor III
X_4 = factor score on Factor IV
X_5 = factor score on Factor V
X_6 = factor score on Factor VI
X_7 = factor score on Factor VII
X_8 = factor score on Factor VIII

The Chi-square test was used in program DSCRIM to determine whether statistically significant differences existed between predictor centroids of the group discriminant scores. The results of this test revealed a significant difference at the 0.05 level ($x^2 = 506.21$; d.f. = 8, $P < 0.0000$).

CALCULATION OF HIT-RATIO

In this case, since the discriminant function is only an *estimate* of whether the respondent answered yes or no to the antisubstitution question, it is necessary to determine the predictive validity of

TABLE 11
STRATIFIED RELIABILITY COEFFICIENTS OF A SCALE FACTORS

Factor	Coefficient (rtt)
I	0.7830
II	0.7053
III	0.8325
IV	0.7834
V	0.7780
VI	0.7080
VII	0.8776
VIII	0.4358
Overall Scale	0.8855

the function. In MDA, the hit-ratio—the percentage of items correctly classified by the discriminant function—is employed for this purpose. As such, it is analogous to R^2 in multiple linear regression. While R^2 is a measure of percent variance explained, the hit-ratio reveals the percentage of correctly classified statistical units.[7]

The computer program ZSCORE, developed by the University of Mississippi Department of Management and Marketing, was used to calculate the hit-ratio. Factor scores of respondents in the hold-out sample plus the discriminant coefficients derived from the analysis sample were used as inputs to calculate a discriminant score for each respondent. A hit-ratio was then calculated based on the "maximum likelihood" criterion for computing the cutting score. This method places the cutting score at the point which is an equal standard deviate from each centroid of the groups.

Program ZSCORE provided a classification matrix denoting the respondents correctly and incorrectly classified. This is presented in Table 12. The classification scheme resulted in a hit-ratio of 87.74%.

To test whether the classification accuracy differed significantly from chance, a Z-test for testing the differences between proportions was performed as outlined by Frank, Massey and Morrison.[14] Using the correction factor for extreme proportions,[15] the difference between the observed (0.88) and expected 0.72) proportions was tested and found significant ($Z_{obs} = 4.12$, $Z_{crit} = 1.96$) at the 0.05 level.

EXAMINATION OF INDEPENDENT VARIABLES

Once the discriminating power of the function was determined to be statistically significant, it was then appropriate to examine the contribution of each independent variable (factor) to the discriminant

TABLE 12
CLASSIFICATION MATRIX FOR HOLD-OUT SAMPLE

Actual Group	Predicted Group		Actual Total	Percent
	A	B		
A	136	17	153	93.79
B	9	50	59	74.63
Predicted Total	145	67	212	
	Hit Ratio = (100) ((136 + 50) ÷ 212) = 87.74%			

Group A = Respondents Favoring Repeal of Antisubstitution Laws
Group B = Respondents Opposing Repeal of Antisubstitution Laws

function. Accordingly, the univariate F-test was used as a test for significant differences between means across each of the eight factors. Table 13 presents results of these univariate F tests as calculated in the DSCRIM computer program. The table indicates that all eight pairs of group means were significantly different on a univariate basis.

Even though all eight factors were statistically significant, such results can be expected when dealing with so many degrees of freedom. To provide a multivariate indication of strength of the relationship between each independent variable (factor) and the discriminant

TABLE 13
UNIVARIATE F-TESTS COMPARING A SCALE FACTOR SCORES
ACROSS TWO GROUPS OF RESPONDENTS

Factor	Group A Mean	Group B Mean	F-ratio	P-Value
I	-0.1226	0.3064	24.6251*	0.0000
I₁	0.0686	-0.2155	10.7916*	0.0015
III	0.1064	-0.3398	26.1754*	0.0000
IV	-0.1336	0.3491	31.7820*	0.0000
V	0.1745	-0.3698	41.0097*	0.0000
VI	-0.0985	0.1309	6.9767*	0.0084
VII	-0.3636	0.9229	335.3150*	0.0000
VIII	-0.0969	0.2362	14.5023*	0.0004

*Significant (α = 0.05, d.f. = 1,633).
Group A = Respondents Favoring Repeal of Antisubstitution Laws
Group B = Respondents Opposing Repeal of Antisubstitution Laws

TABLE 14
CORRELATION BETWEEN EACH INDEPENDENT VARIABLE
AND DISCRIMINANT FUNCTION

Variable (Factor)	Correlation (r)	Coefficient of Determination (r^2)
I	0.2604	0.0608
II	-0.1742	0.0303
III	-0.2682	0.0719
IV	0.2942	0.0865
V	0.3319	0.1102
VI	0.1405	0.0197
VII	0.7919	0.6271
VIII	0.2014	0.0406

scores obtained from the optimally derived discriminant function, DSCRIM provides Pearson product-moment correlation coefficients (r's) between each independent variable and the discriminant scores. Squaring the correlation coefficient provides a measure of the proportion of variance in that variable which is associated with the discriminant score.

Table 14 presents the correlations between each independent variable and the discriminant score. Also included is the coefficient of determination (r^2) with respect to each variable. The table reveals that Factor VII (Pharmacist Autonomy/Essential Pharmacist Services) was, by far, most strongly associated with the discriminant score; 62.71% of its variance was associated with the measure. The percent variance associated between each of the remaining variables and the discriminant score ranged rather smoothly from 1.97 to 11.02%.

Although it has been argued that the antisubstitution controversy is primarily an economic issue, these data do not support such a contention. While Factor IV (Economics of Brand Selection Control) was significantly associated with the respondents' overall response favoring/opposing repeal, only 8.65% of the variance associated with that factor was associated with the discriminant score. Much more strongly associated with the overall attitude favoring/opposing repeal was the factor, Pharmacist Antonomy/Essential Pharmacist Services. Thus, from the pharmacist's point of view, pharmacist autonomy, rather than economics, appears to be the overriding issue in this controversy.

LIMITATIONS OF STUDY

As with all survey research, the investigator must be wary of the potential problem of nonresponse bias. This research is no exception. However, while the usable returns represented a less than optimal proportion of total questionnaires mailed, there is no reason to believe that respondents differed significantly from nonrespondents. Nevertheless, inherent in this investigation is an undefined probability that nonresponse bias did indeed exist. Even with this bias, however, the conclusions with respect to the sample of respondents are not compromised.

REFERENCES

1. What 1,000 Chain Pharmacists Think, *Chain Store Age* 41:26 (Mar.) 1971.

2. Drug Topics Finds Most Rx Men Favor Antisubstitution Laws, *Drug Top.* *114*:47 (July 20) 1970.
3. Antisubstitution Laws, *NARD J. 92*:16 (Oct.) 1970.
4. Strom, B.L., Stolley, P.D., and Brown, T.C.: Drug Anti-Substitution Studies II: Evaluation of Pharmacists' Attitudes Toward Repeal of Anti-Substitution Laws, *Drugs Health Care 1*:104 (Fall) 1974.
5. McCormick, W.C.: Attitudes of Pharmacists, Physicians and Consumers Toward Repeal of Antisubstitution Laws, unpublished Doctor's dissertation, University of Wisconsin, 1972.
6. Nelson, A.A.: The Saliency of Price in the Acceptance of the Pharmacist Substituting Chemically Equivalent Drugs on a Prescription, unpublished Doctor's dissertation, University of Iowa, 1973.
7. Hair, J.F., Anderson, R.E., and Grablowsky, B.J.: An Introduction to Applied Multivariate Statistics, Dynamic Press, Virginia Beach, Virginia, 1974.
8. Guertin, W.H., and Bailey, J.P.: Introduction to Modern Factor Analysis, Edwards Brothers, Ann Arbor, 1970, p. 148.
9. Harman, H.H.: Modern Factor Analysis, University of Chicago Press, Chicago, 1967, p. 311.
10. Harris, C.W.: Some Recent Developments in Factor Analysis, *Educ. and Psych. Meas. 24*:193 (Summer) 1964.
11. Rummel, R.J.: Applied Factor Analysis, Northwestern University Press, Evanston, 1970, p. 392.
12. Veldman, D.J.: Fortran Programming for the Behavioral Sciences, Holt, Rinehart and Winston, New York, 1967, pp. 190–245.
13. Kerlinger, F.N.: Foundations of Behavorial Research, 2nd ed., Holt, Rinehart and Winston, New York, 1973, pp. 447–451.
14. Frank, R.E., Massey, W.F., and Morrison, D.G.: Bias in Multiple Discriminant Analysis, *J. Marketing Res. 2*:250 (Aug.) 1965.
15. Downie, N.M., and Heath, R.W.: Basic Statistical Methods, 2nd ed., Harper and Row, New York, pp. 146–152.

7

Multi-
dimensional
Scaling and
Conjoint
Analysis

CHAPTER REVIEW

This chapter will familiarize you with the benefits of using multi-dimensional scaling (MDS) and conjoint analysis to describe relationships among data. These two techniques are the latest in an increasing family of multivariate techniques that psychometricians and others in the behavioral and statistical sciences have developed to measure and interpret persons' perceptions and preferences. Like other new statistical techniques, their potential is difficult to evaluate at the present stage of development and application. But we are confident that they will receive increasing applications in the future. To more fully understand the concepts, please review the Definitions of Key Terms.

The term "multidimensional scaling" refers to a family of procedures that allow you to show relationships within data with 'pictures' rather than only numbers. Conjoint analysis consists of procedures for estimating the values used by respondents in making judgements or comparisons. After reading this chapter you should be able to:

- [ ] Understand the general concepts of multidimensional scaling.
- [ ] Understand the various types of input that may be used.
- [ ] Understand the basic notion of 'Ideal points.'
- [ ] Be familiar with the basic methods of multidimensional scaling.
- [ ] Understand how one non-metric multidimensional scaling program works.
- [ ] Understand basic approaches to identifying the dimension that may be revealed by MDS.
- [ ] Understand the approach called 'conjoint analysis.'

DEFINITIONS OF KEY TERMS

PREFERENCE DATA.   Indicates the data were gathered by having subjects evaluate stimuli and order them in preference with respect to some property. For example, brand A is preferred over brand C.

SIMILARITIES DATA.   Also referred to as dissimilarities data, it simply implies that subjects evaluate how similar or dissimilar two or more stimuli (objects) are judged to be.

IDEAL POINT.   A hypothetical representation of a stimuli (object or person) possessing just that combination of perceived attributes that are considered "ideal" for the individual.

AGGREGATE/DISAGGREGATE ANALYSIS.   In multidimensional scaling applications aggregate analysis implies that the data that is used in the analysis has been developed by aggregating the responses of many individuals. For example, the responses of 100 people could be examined by using only the average responses (aggregated data). Disaggregated analysis implies that the input data represents individual responses.

MONOTONE. Monotone means having the same order or having a constant direction of change. The series of numbers 1,3,6,9 are said to be monotone because the change from one number to the next is always positive. The series 100,98,70,75,65 is not monotone because the number 75 is not in the right position in the series because the change 70 to 75 is positive while all others are negative.

METRIC/NON-METRIC ANALYSIS. Fully metric analysis assumes that both the input data and the output perceptual configuration are metric. Fully non-metric analysis assumes that the input data are nominal or ordinal (non-metric) and the output configuration is also non-metric. The most frequently used analyses assume that the input data is non-metric while the output perceptual configurations are metric. It is assumed that with sufficient numbers of stimuli the perceptual maps can approximate strongly metric relationships.

7

WHAT IS MULTIDIMENSIONAL SCALING?

"**Multidimensional Scaling**" (MDS) is a family of procedures for transforming unidimensional expressions of relationships into multidimensional expressions of these same relationships. It shows relationships that underlie a unidimensional measure. These various multidimensional representations will usually be characterized by two things: the type of geometric representation used to illustrate the relationships, and the measure of "goodness of fit" of the multidimensional model to the data given by the respondents.

This can be clarified by a simplified illustration in which a HATCO researcher is investigating how respondents perceive six new candybar prototypes. Respondents provided a unidimensional measure of similarity among the candybars by responding to the following questions:

Which pair of candy bars are most similar?
Which pair is next most similar?
(Questioning is continued until all pairs are ordered on their similarity.)

The responses from one person are shown in Table 7.1 and the resulting multidimensional representation might appear as in Figure 7.1.

The unidimensional measures have been transformed into a two-dimensional representation. This representation reveals not only the brands that appear similar or dissimilar, but also enables HATCO to name the axes and identify the basis of the perceptions of similarity. The goals of multidimensional scaling which have been achieved are:

(1) To estimate the underlying dimensions of respondent evaluation, and

TABLE 7.1
ONE RESPONDENT'S JUDGEMENT OF SIMILARITY AMONG
SIX CANDY BARS

Candy Bar	A	B	C	D	E	F
A	–	2	13	4	3	8
B		–	12	6	5	7
C			–	9	10	11
D				–	1	14
E					–	15

(1 = Most Similar Pair)

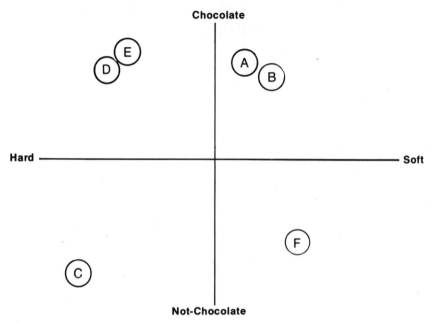

Figure 7.1—Illustration Of Multidimensional "Map" Of Perceptions Of Six Candy Concepts

(2)   To display the objects and/or people on these evaluative dimensions.

To help you understand multidimensional scaling (MDS) procedures we will discuss the basic concepts and assumptions of MDS, the types of data for which MDS provides clarity of understanding, the notion of "ideal points" both from a philosophical and a measurement point of view, and illustrative methods used for MDS.

BASIC CONCEPTS AND ASSUMPTIONS OF MDS

Any phenomenon (product, service, image, aroma, etc.) can be thought of as having both perceived dimensions and objective dimensions. For example, HATCO management may see their product (a lawnmower) as having two color options, a 2 h.p. motor, and a 24 inch blade. These would be called the objective dimensions of the product. On the other hand, customers may (or may not) see these attributes. But they may also perceive the HATCO mower to be "expensive looking" and "fragile." These are examples of perceived dimensions. Two products having the same physical characteristics (objective dimensions) but bearing different brand insignia may be perceived to differ in quality (a perceived dimension) by many customers. This phenomenon is known to business people as well as social scientists and recognition of it must be a part of business decision-making. Two features of the differences between objective and perceptual dimensions are very important:

a. The dimensions perceived by customers may not coincide with (or may not even include) the objective dimensions assumed by the researcher.

b. The evaluations of the dimensions (even if the perceived dimensions are the same as objective dimensions) may not be independent and may not agree. For example, one soft drink may be judged sweeter than a second soft drink because the first has a "fruitier" aroma although both contain the same physical amount of sugar.

Multidimensional scaling techniques enable the researcher to represent respondents' perceptions spatially. That is, to create visual displays that represent the dimensions perceived by the respondents when evaluating stimuli (e:g: brands, objects etc.). This visual representation helps the researcher better understand similarities and dissimilarities between objective and perceptual dimensions. These visual representations or "pictures" are often referred to as spatial maps (as illustrated in Figure 7.1). Spatial maps have not been directly proven to represent perception but have provided insights into perception. We may also assume (although not prove) that all respondents will not:

a. Perceive a stimulus to have the same dimensionality (although it is thought that most people judge in terms of a limited number of characteristics or dimensions). For example, some people might evaluate a car in terms of its horsepower, while others do not consider horsepower at all when evaluating an automobile.

b. Attach the same level of importance to a dimension even if all respondents perceive this dimension. For example, two people may both perceive a cola in terms of its level of carbonation but one considers it unimportant while the other considers this dimension (carbonation) to be very important.

c. Judge a stimulus in terms of either dimensions or levels of importance that remain stable over time. In other words, one may not expect people to maintain the same perceptions for long periods of time.

In spite of any weaknesses caused by these assumptions, we will attempt to represent perceptions spatially such that any underlying relationships might be examined.

We will limit our discussion to the techniques of multidimensional scaling (MDS) and touch upon the psychological underpinnings only when necessary to qualify a point. You should be cautioned in two areas:

a. MDS computer programs are becoming readily available to those having access to a computer. The possibilities for abuses of MDS are great because of the complex theory underlying the measurement of perception and the assumed relationships among the dimensions of perception, both of which must be understood before using and interpreting MDS results.

b. Programs now available are based on many different rationale for scaling. Each program must be carefully scrutinized for the assumptions underlying its successful use.

The variety of available computer programs for MDS is rapidly expanding. We will use an illustrative program but, through concentration on the types of input data, the desired types of spatial representations, and interpretational alternatives, we will provide an overview of MDS that will allow you to readily understand the nature of individual program differences.

INPUT TO MDS

When individual perceptions of stimuli are obtained from respondents, a wide variety of techniques or procedures may be used, but the resultant data may be generally categorized as preference data, similarities data (proximity data), and ideal points. These data serve as the basis for using MDS techniques to derive the perceptual dimensions used by the respondents to judge the similarity of, preference for, and/or ideal stimuli.[1]

Preference Data

Preference implies that stimuli be judged in terms of dominance relationships; that is, the stimuli are ordered in preference with respect to some property. For example, Brand A is preferred over Brand

---

[1] A good deal of the discussion in this chapter is based on the program descriptions available with the many M.D.S. programs and the works of Green with Rao, Carmone, and Robinson [2, 3, 4, 5]. When specific reference is made it is properly footnoted. It would be difficult (impossible?) to write on M.D.S. without drawing directly from these sources. If similarity in presentation is noted, it is acknowledged here.

C. For *n* subjects and k stimuli, one could ask the subjects to rank order the stimuli with the rank "1" assigned to the most preferred. The result would be an *n* x k matrix of subjects and stimuli in which comparisons are possible within rows only. That is, data is seldom available in which comparison of two rows (subjects) is valid. One could determine that subject 1 prefers stimulus A to stimulus B, but it would be difficult to determine if subject 1 prefers stimulus A more or less than subject 2 perfers stimulus A, since there is no apparent basis for comparing across subjects.

Preference data is also commonly gathered by presenting paired comparisons to many subjects and then using *aggregated* values (for example, utilizing Thurston's Law of Comparative Judgment). In this situation there is only one set of variables—the rank ordered preferences for stimuli. This aggregation of paired comparisons data often presents problems of intransitivity (e.g., $A>B>D>A$) and inconsistencies among the rankings of the several subjects. After testing for inconsistencies (using, for example, Kendall's coefficient of concordance) a procedure has been proposed by Carmone, Green, and Robinson [2] to resolve some of the intransitivity problems.

Another procedure for gathering preference data is to rate each stimuli on an explicit scale, such as a numerical ten point scale, or a scale showing relative distance from an anchor point. These scales assume the subject is capable of rating all stimuli on the same scale. The above mentioned procedures are not exhaustive of the possible means of obtaining preference data, but should serve as a guide to the reader.

Similarities (Dissimilarities) Data

Several procedures are available for gaining respondents' perceptions of the similarities (or dissimilarities) among stimuli. A few illustrative procedures for gathering similarities data are presented; again, no pretense at an exhaustive list is made. The examples presented here include procedures for obtaining disjoint data, conjoint data, confusion data, and derived measures of similarity (or dissimilarity) [3].

Conjoint and Disjoint Measurement. Often the terms conjoint and disjoint describe data used for multidimensional scaling input. A simple interpretation is:

> Conjoint data: Data from sets of stimuli in which the sets share a common member.
>
> Disjoint data: Data from sets of stimuli in which the sets **do not** share a common member.

For illustration we will look at all possible pairs of four brands

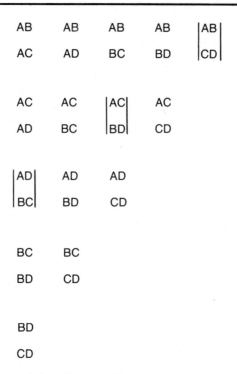

Figure 7.2—All Pairs Of Pairs For Four Brands

of soft drink (Brands, A, B, C, and D) in Figure 7.2. Those pairings enclosed in brackets are disjoint because they do not share a common member. Pairings (sets) not in brackets are examples of conjoint data because they share a common member.

   With four brands we have six distinct pairs of brands $((n(n-1))\div 2)$ and 15 possible pairs of pairs $(\frac{1}{8}(n+1)(n)(n-1)(n-2))$. These 15 form all conjoint and disjoint comparisons between pairs of pairs. There are 12 $(\frac{1}{2}(n(n-1)(n-2)))$ pairs of conjoint measures and three pairs of disjoint measures. If we obtained only the conjoint or disjoint set, we would have to assume relationships measured by those comparisons we did not obtain.

   Examples of Conjoint Data.   Data obtained from comparisons among groups so that each group has a member in common with another group are termed conjoint data. To obtain conjoint data:

   a.   Choose from each triad of stimuli (do this for all possible triples) that pair which is most similar and that pair most dissimilar. This can lead to a large number of comparisons if many stimuli are involved. Four stimuli would have four triads, five stimuli would have ten. The formula for calculating the number of triads is:

$$\frac{n!}{(n-3!)\ (3!)}$$

The task could quickly become unmanageable for the respondents.

Example: Select from ABC

Pair Most Similar	A	B
Pair Least Similar	A	C

b. Rank each of the $n-1$ stimuli from a reference stimulus (closest is most similar to reference) with each stimulus in turn serving as the stimulus to which the others are compared. This produces a matrix for each subject.
Example:

		Reference		
		A	B	C
	A	–	1	2
Stimuli	B	1	–	2
	C	2	1	–

Row 1 in the matrix indicates that stimulus B is more similar to stimulus A than is stimulus C, and Row 3 indicates that stimulus B is more similar to stimulus C than is stimulus A.

Conjoint and Disjoint Data

Many times researchers do not wish to infer any of the order relationships among their data but would prefer to measure them directly. This calls for a comparison of all pairs of data—both conjoint and disjoint pairs, as indicated below.

a. Each respondent is given all possible pairs, as in Figure 7.2, and asked to select the most similar pair from each pair of pairs. This data would produce a matrix for each respondent that orders the similarities of all possible comparisons.

b. Rank all distinct pairs of stimuli according to relative similarity. If we have stimuli A, B, C, D, and E, we could rank pairs AB, AC, AD, AE, BC, BD, BE, CD, CE, and DE, in the order from that pair which is most similar to that pair in which the members are least similar. If, for example, pair AB were given the rank of 1, we would assume that the respondent sees that pair as containing the two stimuli that are most similar, as contrasted to all other pairs.

An illustrative approach would be to cluster stimuli into groups such that stimuli within a group are more like each other than like those in other groups.

Example: Group 1   Group 2
                     AB   EFC

(this can be repeated to refine the measure).

The data can be treated as confusion matrix (discussed in following section), by recording a 1 if two stimuli appear in the same group and a zero if they do not. The data would typically be aggregated across respondents.

Confusion Data. The pairing (or "confusing") of stimuli i with stimuli j is taken to indicate similarity. For example: respondents are given a number of soft drinks and branded descriptions of the soft drinks and asked to assign the soft drinks to the descriptions. Through either replication to obtain a confusion matrix (a matrix whose cells contain the incidence of pairing stimuli i with stimuli j) for an individual or by pooling respondents' judgments (you must assume homogeneous perception of dimensions), a matrix of confusion scores is obtained.

For example, a confusion matrix representing 4 respondents' perceptions of three soft drinks (A, B, and C) might appear as:

		Description		
		A	B	C
	A	2	1	1
Stimulus (Brands)	B	1	1	2
	C	0	2	2

This illustration reveals that of four respondents, one confused Brand A with Brand B and one confused Brand A with Brand C (by examination of Row 1).

Derived Measures. Derived similarities are typically based on "scores" given to stimuli by respondents. For example: subjects are asked to evaluate k stimuli on two semantic differential scales. For example: rate cherry soda, strawberry soda, and a lemon-lime drink on the following scales:

Sweet   ＿＿＿ ＿＿＿ ＿＿＿ ＿＿＿ ＿＿＿ ＿＿＿Tart
Light Tasting ＿＿＿ ＿＿＿ ＿＿＿ ＿＿＿ ＿＿＿ ＿＿＿Heavy

The 2x3 matrix could be evaluated for each respondent (correlation, agreement, etc.) to create similarity measures.

Two important assumptions are involved here:

   a. That the researhcer has selected the appropriate dimensions to measure with the semantic differential, and

   b. That the scales should be weighted (either equally or unequally) to achieve the similarities data for a subject or group of subjects.

In summary, these procedures (Preference, Similarity, etc.) have a common purpose of obtaining a series of unidimensional responses that represent the respondents' judgments so that they define the underlying multidimensional pattern leading to these judgments. They serve as inputs to the many procedures labeled multidimensional scaling.

Ideal Points

We can assume that if we locate (in geometric space) that point which represents the most **preferred combination** of perceived attributes (on all relevant attribute dimensions), we would have a spatial representation of a subject's "ideal" stimuli. Equally, we will assume that the position of this ideal point (relative to other stimuli on our derived spatial map) would tend to define the **relative** preference of all other stimuli. The determination of the spatial location of the ideal point and the saliences of the relevant dimensions usually evolves from two approaches: explicit estimation or implicit estimation.

Explicit estimation proceeds from the direct responses of subjects. This could involve asking the subject to rate an hypothesized ideal on the same attributes that the other stimuli are rated, or the respondent is asked to include among the stimuli used to gather similarities data an hypothesized ideal stimuli (brand, image, etc.).

When asked to conceptualize an "ideal" of anything we typically run into problems. Often, the respondent will simply conceptualize the ideal at the extremes of the explicit ratings used, or conceptualize the ideal to be similar to the most preferred product from among those with which the respondent has had experience. Often these perceptual problems lead the researcher to implicit ideal point estimation.

There are several procedures for implicitly positioning ideal points. The basic assumption underlying most procedures is that derived measures of ideal points' spatial positions are maximally consistent with individual respondent's provided perceptions (usually preferences).

Differing concepts of ideal point estimation assume:

a. Ideal points for all respondents share common dimensionality and weighting of dimensions.
b. Weighting of the axes idiosyncratically for respondents. For

example, two respondents may judge a soft drink identically on aroma, but aroma is unimportant to one while very important to another.

c. Idiosyncratic axis rotation and weighting although the dimensions are common to all respondents.

Srinivasan and Shocker [9] assume an ideal point can be derived as follows: the ideal point for all pairs of stimuli is determined so that it violates with least harm the constraint that it be closer to the most preferred in each pair than it is to the least preferred in each pair. For example, if A is preferred to B and C is preferred to D, the ideal point should be closer to A than to B while also constrained to be closer to C than to D.

In summary, there are many ways to approach ideal point estimation, and no one has been demonstrated to be a "best" method. The choice of the ideal point estimation procedure will depend on the researcher's skills and the specific MDS procedure selected for use.

METHODS OF MULTIDIMENSIONAL SCALING

The various approaches to multidimensional scaling can be categorized regarding:
1. level of measurement assessed for input and output.
     a. fully non-metric
     b. non-metric
     c. metric
2. aggregate vs. disaggregate (for similarities data)
3. internal vs. external analysis (for preference data).
The following sections will discuss and illustrate these points.

Fully Non-Metric Methods

Fully non-metric methods assume both ordinal input and ordinal output. This type of scaling is easiest to visualize in the unidimensional case. Figure 7.3a illustrates an "unfolding" model for preference data from two subjects.

The points A and B on the line indicate the ideal points for subjects A and B. By "folding" the scale you can get the ordering of their preferences for stimuli products 1, 2, 3, and 4 (as indicated in the figure). Imagine folding the scale at the point marked "A". You would then have person A's order of preference as shown in Figure 7.3b. It would seem that if asked, respondent A would say that she prefers the products (stimuli) in the order 2, 1, 3, 4. However, we

a. Aggregate Perception of the stimuli

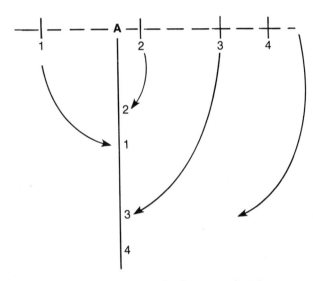

b. Preference order for respondent A

Figure 7.3—Illustration of Unfolding Model for Preference Data From Two Subjects

seldom know the real perceptions of the stimuli as in our example. What really happens is that we get preference data from subjects and use it to derive the aggregate preference scale. For example, suppose respondents A and B indicate the preference orders below:

A:	2	1	3	4
B:	3	4	2	1.

If we get other respondents with preference orderings consistent with these we might find that we can "unfold" these preferences as shown in Figure 7.4. This gives us a view of the relative differences in the stimuli and where the subject's ideal points lie on the stimuli preference scale.

The model is based on the assumption that the rank order of preference for a subject corresponds to the rank order of the distances of the stimuli from the position of the subject's ideal point

Aggregate preference data →

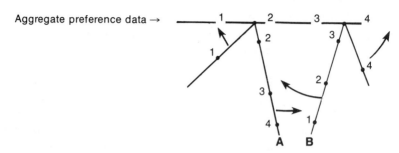

Figure 7.4—Illustration of Unfolding Model for Aggregate Preference Data

on the underlying continuum. This has been generalized to the multi-dimensional case [4] in which the distances from the ideal points to the stimuli points represent the ordering of preferences on each dimension.

Non-Metric Methods

Non-metric methods assume ordinal input and metric output, i.e., the distances output by the MDS procedure may be assumed to be at least intervally scaled. The criterion for determining the location of stimuli (and/or subjects) in t-dimensional (where t is any dimensionality from 1 to the number of stimuli minus 1) space is to find configurations (output spatial maps) whose rank orders of estimated ratio scaled distances between all stimuli best reproduces the input rank orders. For each level of dimensionability from t-dimensions to 1 dimensions a "stress' measure is calculated, and one attempts to find the lowest level of dimensionality ((you can usually understand a 2 dimensional spatial map with far greater facility than a 3 dimensional map) producing satisfactory "stress." For example, with data available for the paired dissimilarities among five advertisements we would have ten pairs of inequality measures. If we arbitrarily place five points in two dimensional space we could then move these five points until the distances between them (remember the calculation is now based on two dimensions) violated the ten inequalities as little as possible. If we are using the Euclidean distance between the points, the configuration achieved in this fashion will have the same monotone (rank) distance as the inequalities. You could add a constant to all dimensions, rotate the axes, change to the mirror-image of the original configuration, or uniformly stretch or compress the scales of the dimensions and not change the monotone relation. The major criteria is preservation of the monotone relation between the original inequalities and the distances between points. The "stress' measure mentioned is simply a measure of how well (or poorly) the ranked distances on the map agree with the ranks given by the respondents.

Illustration. HATCO is contemplating entry into the softdrink market. Uncertain of what type of softdrink to produce for maximum perceived consumer satisfaction, HATCO management took a random sample of consumers and elicited their preferences among six soft-drinks currently on the market (in identical containers with no brand names) and the consumers' judgments on the similarities of these same softdrinks. Figure 7.5 illustrates a two-dimensional map derived from one persons' perceptions of six softdrinks.

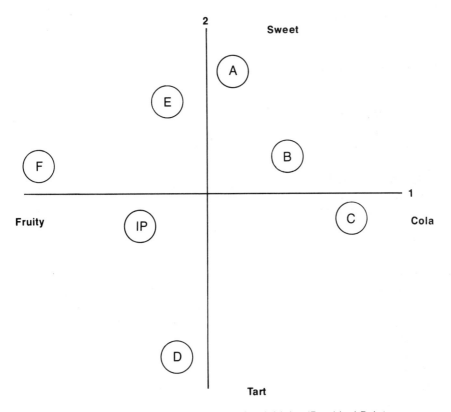

KEY: A, B, C, D, E, F Indicate the 6 softdrinks; IP + Ideal Point

Figure 7.5—Two-dimensional Map Of A Single Subject's Views Of the Simi-larities Among 6 softdrinks

(1 equals most similar pair and 15 equals least similar pair)

Softdrink	A	B	C	D	E	F
A	—	3	6	15	1	8
B		—	2	12	4	9
C			—	10	7	13
D				—	14	11
E					—	5
F						1

Figure 7.6—Six softdrinks ranked by similarity

The axes were labeled as follows: axis I—fruit flavored drinks to cola drinks, axis II—sweet tasting to tart tasting. Drinks A-B-C were colas and D-E-F were lemon-lime, orange, and strawberry flavored drinks. The various approaches to labeling will be discussed later. In this spatial mapping of similarities the distances among all the points preserve (as well as possible) the rank-order of the similarities judged by the respondent. That is, if we had a matrix such as that in Figure 7.6 which revealed the ranked similarities between the softdrinks we could take all of the straight line distances between all of the points on the map (Figure 7.5) and they should rank in the same order as this similarities matrix.

Because we are trying to locate six points in two-dimensional space so that they preserve these 15 relative evaluations, the configuration we derive should give an indication of the dimensions the respondent used in judging the softdrinks. We will assume that we have also obtained a spatial representation of the respondents' "ideal" softdrink—identified by point IP in Figure 7.5. If we can assume these representative perceptions relate to choice we might conclude that this subject would select from among these drinks in order F, E, D, B, A, C. This ordering represents the ordering of the straight-line (Euclidean) distances from point IP to all the points representing the softdrinks. We are assuming the direction of difference is not critical—only the absolute distance. Additionally, this respondent's ideal softdrink would be a slightly tart drink perceived as more fruity than cola. This assumption is often not borne out by further experimentation. For example, Raymond [8] cites an example in which the conclusion was drawn that people would prefer brownie pastries on the basis of degree of moistness and chocolate content. When the food technicians applied this result in the laboratory they found that their brownies made to the experimental specification became chocolate milk. One cannot always assume the relationships found are independent, linear or hold over time as noted previously. However, multidimensional scaling is a beginning in understanding perceptions and choice that will expand considerably in the next few years.

Metric Methods

Metric methods assume that input as well as output data are metric. This assumption allows us to strengthen the relationship between the final output dimensionality and the input data. Rather than assuming we preserve only the monotone relationship, we can assume that the output preserves the interval and/or the ratio of the input data. While this assumption is often hard to support with the data available to marketing researchers, the results of non-metric and metric procedures applied to the same data are often very similar.

Aggregate vs. Disaggregate Analysis

For similarities (dissimilarities) data we are dealing with perceptions of stimuli on a proximity basis of measurement and the output consists of representations of the stimuli proximities in t-dimensional space (when it is less than the number of stimuli). The researcher can generate this output on a subject-by-subject basis (producing as many "maps" as subjects) or attempt to create fewer maps by some process of aggregation.

The aggregation may take place either before or after scaling the subjects' data. Before scaling, the researcher might cluster analyze the subjects' responses to find a few "average" or representative subjects, or just pool the data and use the average response. This latter method should be viewed with caution as it assumes commonality of dimensions and saliences across all subjects—potentially a dangerous assumption. After scaling, the resulting maps could be compared (cluster analysis might be used for this comparison also) to determine if groups of subjects have commonality of perception. Neither approach is inherently better than the other and the cautious researcher would be wise to examine both approaches.

Internal vs. External Analysis of Preference Data

Internal analysis of preference data refers to the development of a spatial map shared by both stimuli and subject points (or vectors) solely from the preference or dominance data. As an example of the flexibility available, Kruskal and Carmone's M-D-Scal program [7] allows the user to find configurations of stimuli and ideal points assuming no difference between subjects at all, assuming separate configurations for each subject, or a single configuration with individual ideal points. By gathering preference data only the researcher can still represent both stimuli and respondents in a spatial map.

External analysis refers to fitting ideal points (based on preference data) to stimulus space developed from similarities data obtained from the same subjects. For example we might individually scale similarities data, examine the individual maps for commonality of perception, and then scale the preference data for any groups identified in this fashion. This means the researcher has to gather both preference data and similarities data to achieve external analysis.

Green and Rao [4] hold the view that external analysis is clearly preferable in most instances. Their preference is due both to computational difficulties with internal analysis procedures as well as the confounding of differences in preference with differences in perception. Additionally, the saliences of perceived dimensions may change as one moves from perceptual space (are the stimuli similar or dissimilar?) to evaluative space (which stimuli is preferred?).

A GENERALIZED MDS APPROACH

To illustrate an approach to MDS, we will present the simplified sequence of steps in multidimensional scaling from a commonly available non-metric program: M-D-Scal [7].

Step 1. Select some initial configuration of stimuli (s_k) at a desired initial dimensionality (t). M-D-Scal uses either a configuration supplied by the researcher based on previous data, or generates its own configuration by selecting pseudo-random points from an approximately normal multivariate distribution.

Step 2. Compute the distances between the stimuli points, evaluate the index of fit and/or evaluate the "stress" of the configuration. M-D-Scal computes the interpoint distances (dij) in the starting configuration and (for metric data) performs a least squares regression of d*ij* on the original data distances (S*ij*). The estimated d*ij* values termed dij are used to calculate the following stress measure:

$$\text{Stress} = \frac{\Sigma(\text{dij} - \text{dij})^2}{\Sigma(\text{dij} - \text{dij})^2}^{1/2}$$

where dij is the average distance (Σdij/n).

Step 3. If the stress measure is greater than some small stopping value selected by the researcher—find a new configuration where the stress is further minimized. M-D-Scal uses the method of steepest descent to find a new configuration. This essentially involves evaluating the partial derivatives of the stress function to determine the directions in which the best improvement in stress is to be obtained, and moving the points in the configuration in those directions in small increments.

Step 4. The new configurations are evaluated and adjusted until satisfactory stress is achieved.

Step 5. Once satisfactory stress has been achieved, the dimensionality is reduced by 1 and the process is repeated until the lowest dimensionality with acceptable stress has been reached.

Identifying the Dimensions

As we discussed briefly in the Factor Analysis chapter, identifying underlying dimensions is often a difficult task. Multidimensional scaling techniques have no built-in procedure for labeling the dimensions. The researcher, having developed the "maps" in the selected dimensionality, can adopt several procedures:

1. Respondents may be asked to subjectively interpret the dimensionality by inspecting the "maps."
2. The researcher may identify the axes in terms of objective characteristics of the stimuli.
3. If the similarities (or preference) data were obtained directly, the respondents may be asked (after stating the similarities and/or preferences) to identify the characteristics most important to them in stating these values. The set of characteristics can then be screened for values that "match" the relationships portrayed in the maps.

4. The subjects may be asked to evaluate the stimuli on the basis of researcher-determined criteria (usually both objective values) and researcher perceived subjective values. These evaluations can be compared to the stimuli distances on a dimension-by-dimension basis for labeling the dimensions.

These procedures, while not exhaustive of those suggested, reveal the difficulty of labeling. This task is one that cannot be left unattended until the multidimensional scaling procedures are implemented. The researcher must plan the use of alternative labeling schemes early in the design of the research.

CONJOINT MEASUREMENT

Frequently a researcher is faced with determining the best set of attributes for a product or service. For example, suppose you could build apartments with 1, 2, or 3 bedrooms, and with 1, 2, or 3 baths, to rent for $300, $350, and $400 a month. What should you build? The problem would be simple if you knew the values (or 'utilities') placed on each of these attributes by your prospective tenants.

Let's look at an example. If we knew the relative values to be as follows:

Bedrooms		Bathrooms		Rent	
Number	Value	Number	Value	$	Value
1	−2.1	1	−1.5	300	2
2	1.5	2	.9	350	1
3	.6	3	.6	400	−3

We could quickly determine that:
1) rent is the most important feature because a change in rent from $300 to $400 causes a change in utility from 2 to −3; a 5 unit change. This is larger than the change in utility over the range of any other feature of the apartment.
2) the 'best' combination would be found by adding the highest values for each attribute:

highest utiliy = 2 bedrooms + 2 bathrooms + $300 rent
    4.4   =   1.5   .9   2

All combinations of these three attributes can be examined for relative total value. For illustration, the *poorest selection* would be found by adding all of the lowest values for each atribute:

Lowest value = 1 bedroom + bath + $400 rent
−6.6   =   −2.1   1.5 −   3

The issue to be addressed in the following section is: How do we find these values? To answer this question we will discuss two techniques: Monotone regression using a Factorial design and Trade-off analysis.

Monotone Regression

To find the values for the attributes of our proposed apartment project we could treat the problem as a factorial experiment. We could probably best visualize the problem in the form of an experiment as shown in Figure 7.6. With 3 attributes and each having 3 levels we could have 27 different apartments that we could build, such as a 1 Bedroom, 1 bath for $300 or a 2 Bedroom, 1 bath, for $400.

To approximate the values people place on each attribute we could give prospective tenants (in random order) cards printed with each of the 27 descriptions and have them rank order the cards in terms of the desirability of the apartments described. Figure 7.7 shows the ranks given by one respondent. A computer program called Monanova would be used to find the values that when attached to each level attribute would add together to reproduce the rank order given by this respondent. Since the purpose is to produce coefficients for each level of each attribute that add to scores that have only the **same rank order** as the original data, it is called mono-

Rent

		$30	$350	$400
1 Bedroom	1 Bath	19	23	27
	2 Baths	12	15	25
	3 Baths	13	16	26
2 Bedrooms	1 Bath	7	11	23
	2 Baths	1	4	17
	3 Baths	2	6	18
3 Bedrooms	1 Bath	10	14	24
	2 Baths	3	8	20
	3 Baths	5	9	21

Figure 7.7—Desirabilities Rankings For Apartments

tone regression. Monotone refers **to having** a constant direction of change (order). For example, the two sets of observations below:

A	1	2	3	4
B	.3	.4	.9	1.11

are said to be monotonic because the **order** of both sets is identical. That is, the direction of change from 1 to 4 and .3 to 1.11 is the same (higher) over every interval from both sets of observations. If we assume data set A to be the ranks for 4 apartments given by a respondent we would try to produce values for the attributes (bedrooms, bathrooms, rent) whose sums would be perfectly monotone (same rank order) with the original respondent ranking for all apartments. You can think of this process as fitting a dummy variable regression model as described in Chapter Two.

One drawback to gathering this type of data is the difficulty of having people rank many items (in this case 27). However, we can use less than full ranking through fractional factorial designs [5].

Another method has been devised for producing the attribute values. This method is called trade-off analysis.

Trade-Off Analysis

Instead of presenting our prospective tenants with 27 apartment concepts, we present them with 3 matrices as shown below: (shown as completed by a respondent)

Bedrooms				Rent				Bathrooms		
Bathrooms				Bedrooms				Rent		
	1	2	3		300	350	400		1	2	3
1	9	5	6	1	5	6	9	300	5	1	2
2	7	1	3	2	1	3	7	350	6	3	4
3	8	2	4	3	2	4	8	400	9	7	8

Each respondent is asked to rank each combination in each matrix from 1 to 9 in order of desirability. You can see that our hypothetical respondent would prefer 2 bedrooms and 2 baths over any other combination of bedrooms and bathrooms. Although every attribute is ranked against the others it is assumed that the respondents are capable of considering all attributes (rent, bedrooms, and bathrooms) independently of the others.

By making this assumption you are able to avoid having each concept evaluated on *all* attributes simultaneously as we did in the factorial design.

Once the respondents have ranked all 3 sets of matrices (called trading-off the attributes) we can again create a 'monotone' additive model to describe the values. In this situation exactly the same values will be created. For both of the examples you can quickly check this with pencil and paper calculations.

An essential difference between the experimental design approach and the trade-off approach is that:

a. in the experimental design approach the respondent gives preference for all of the various apartments and we are concerned with estimating the relative influence of the various attributes of an apartment on this choice.

b. in the trade-off approach the respondents gave only a preference for combinations of features of apartments—not the apartment as a total set of attributes. We infered the 'best' apartments combinations by summing the highest weights after estimating the relative influence of the attributes of an apartment.

Application Precautions

Both of the approaches described above assume that the attributes evaluated by respondents *do not* interact in creating an impression of desirability. For example, we assume that a change from one bathroom to two bathrooms has the same effect on choice when the apartment has one bedroom or two. This assumption is often difficult to support. Using conjoint analysis will give poor (and often misleading) results if the data interact. Note also the assumption that the relative utilities of the measured attributes (bathrooms, bedrooms, and rent) can be described with a simple additive model. This assumption may not always hold and it is difficult to detect when it does not hold.

What Have We Learned?

The following represent what conjoint analysts can tell us:

☐ The relative importance of each of the attributes of an apartment.

☐ The values placed on each level of each attribute in so far as they influence the preference for apartments.

☐ Which apartment is best and in general how sensitive the selection is to changing the attributes.

The techniques of conjoint analysis also are very valuable to the researcher when:

1. Attempting to determine the relative values people attach to the attributes of stimuli that appear to influence their choices.

2. Predicting changes in preference from the attribute values.

SUMMARY

In this chapter we presented an overview of the objectives, input measures, assumptions, and output for illustrative cases of multi-dimensional scaling and conjoint analysis procedures. You are not expected to be immediately capable of implementing these procedures. But should now be prepared to examine critically the assumptions and procedures of MDS and conjoint analysis procedures when their use is contemplated. Multi-dimensional scaling and conjoint analysis face the same kinds of limitations that confront any type of survey, or laboratory-like, technique. While some successful applications have been reported, the number is still too small to establish a convincing track record of their widespread usefulness at the present time.

END OF CHAPTER QUESTIONS

1. Suppose you wished to see how your classmates view six of your professors at your college. Your purpose is to attempt to identify the differences or similarities among the professors and to specify the bases of the similarities or differences. Could you suggest how you might obtain the appropriate data?
2. Can you describe a problem for which multidimensional scaling might be appropriate? What characteristics of the problem make the technique of multidimensional scaling appropriate?
3. Would you attempt to assign an "statistical" meaning to the results obtained from the application of conjoint analysis?
4. What advantages and disadvantages can you identify when contrasting the two approaches to conjoint analysis presented in this chapter?
5. What advantages or disadvantages do you see in basing a multidimensional scaling procedure on explicitly stated dimensions of comparison for stimuli versus inferring the dimensions from the direct comparison of the stimuli only?

REFERENCES

1. Bennett, J. F. and W. L. Hays, "Multidimensional Unfolding: Determining the Dimensionality of Ranked Preference Data", *Psychometrika*, U. 25 (1960), 27-43.
2. Carmone, F. J., P. E. Green, and P. J. Robinson, "Tricon—an IBM 360-65

       FORTRAN IV Program for the Triangulatization of Conjoint Data".
       Journal of Marketing Research, Vol. 5 (May, 1968) 219-20.
3. Green, P. E. and Y. Wind, Multiattribute Decisions in Marketing, Dryden
       Press (1973), pp. 72-73.
4. Green P. E. and V. Rao, *Applied Multidimensional Scaling: A Compari-*
       son of Approaches and Algorithms. New York: Holt, Rinehart & Winston,
       1972.
5. Green, P. E. and F. J. Carmone, *Multidimensional Scaling and Related*
       Techniques in Marketing Analyses, Allyn and Bacon (1970).
6. Johnson, R. M. "Trade Off Analysis of Consumer Values", *Journal of*
       Marketing Research, Vol. II (May, 1974) 121-7.
7. Kruskal, J. B. and F. J. Carmone. "How to Use M-D-Scal (version 5M)
       and Other Useful Information", Bell Telephone Laboratories, Murray
       Hill, N.J., March 1969.
8. Raymond, Charles, *The Art of Using Science in Marketing.* Harper and
       Row (1974) pp. 91-92.
9. Srinivasan, V. and A. D. Shocker, "Linear Programming Techniques for
       Multidimensional Analyses of Preferences", *Psychometrika* 38 (Sep-
       tember, 1973).

SELECTED READINGS

MARKET SEGMENTATION:
A STRATEGIC MANAGEMENT TOOL
RICHARD M. JOHNSON*

    Like motivation research in the late 1950's, market segmentation
is receiving much attention in research circles. Although this term
evokes the idea of cutting up a market into little pieces, the real role
of such research is more basic and potentially more valuable. In this
discussion *market segmentation analysis* refers to examination of the
structure of a market as perceived by consumers, preferably using a
geometric spatial model, and to forecasting the intensity of demand
for a potential product positioned anywhere in the space.
    The purpose of such a study, as seen by a marketing manager,
might be:

    1. To learn how the brands or products in a class are perceived
       with respect to strengths, weaknesses, similarities, etc.
    2. To learn about consumers' desires, and how these are satis-
       fied by the current market.
    3. To integrate these findings strategically, determining the great-

    * Richard M. Johnson is Vice President of Market Facts, Incorporated.
Reprinted by permission of the authors and publisher from the Journal of Marketing
Research, *published by the American Marketing Association Vol. viii (February 1971)
pp. 13–18.*

est opportunities for new brands or products and how a product or its image should be modified to produce the greatest sales gain.

From the position of a marketing research technician, each of these three goals translates into a separate technical problem:

1. To construct a product space, a geometric representation of consumers' perceptions of products or brands in a category.
2. To obtain a density distribution by positioning consumers' ideal points in the same space.
3. To construct a model which predicts preferences of groups of consumers toward new or modified products.

This discussion will focus on each of these three problems in turn, suggesting solutions now available. Solutions to the first two problems can be illustrated with actual data, although currently solutions for the third problem are more tentative. This will not be an exhaustive catalog of techniques, nor is this the only way of structuring the general problem of forecasting consumer demand for new or modified products.

CONSTRUCTING THE PRODUCT SPACE

A spatial representation or map of a product category provides the foundation on which other aspects of the solution are built. Many equally useful techniques are available for constructing product spaces which require different assumptions and possess different properties. The following is a list of useful properties of product spaces which may be used to evaluate alternative techniques:

1. *Metric:* distances between products in space should relate to perceived similarity between them.
2. *Identification:* directions in the space should correspond to identified product attributes.
3. *Uniqueness/reliability:* similar procedures applied to similar data should yield similar answers.
4. *Robustness/foolproofness:* procedures should work every time. It should not be necessary to switch techniques or make basic changes in order to cope with each new set of data.
5. *Freedom from improper assumptions:* other things being equal, a procedure that requires fewer assumptions is preferred.

One basic distinction has to do with the kinds of data to be analyzed. Three kinds of data are frequently used.

Similarity/Dissimilarity Data

Here a respondent is not concerned in any obvious way with dimensions or attributes which describe the products judged. He makes global judgments of relative similarity among products, with the theoretical advantage that there is no burden on the researcher to determine in advance the important attributes or dimensions within a product category. Examples of such data might be: (1) to present triples of products and ask which two are most or least similar, (2) to present pairs of products and ask which pair is most similar, or (3) to rank order k-1 products in terms of similarity with the kth.

Preference Data

Preference data can be used to construct a product space, given assumptions relating preference to distances. For instance, a frequent assumption is that an individual has ideal points in the same space and that product preference is related in some systematic way to distances from his ideal points to his perception of products' locations. As with similarity/dissimilarity data, preference data place no burden on the researcher to determine salient product attributes in advance. Examples of preference data which might lead to a product space are: (1) paired comparison data, (2) rank orders of preference, or (3) generalized overall ratings (as on a 1 to 9 scale).

Attribute Data

If the resarcher knows in advance important product attributes by which consumers discriminate among products, or with which they form preferences, then he may ask respondents to describe products on scales relating to each attribute. For instance, they may use rating scales describing brands of beer with respect to price vs. quality, heaviness vs. lightness, or smoothness vs. bitterness.

In addition to these three kinds of data, *procedures* can be *metric* or *nonmetric*. Metric procedures make assumptions about the properties of data, as when in computing a mean one assumes that the difference between ratings of values one and two is the same as that between two and three, etc. Nonmetric procedures make fewer assumptions about the nature of the data; these are usually techniques in which the only operations on data are comparisons such as "greater than" or "less than." Nonmetric procedures are typically used with data from rank order or paired comparison methods.

Another issue is whether or not a *single product space* will adequately represent all respondents' perceptions. At the extreme, each

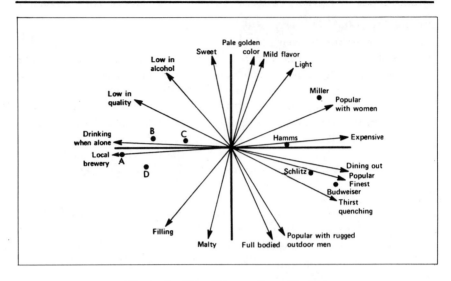

Figure 1—The Chicago Beer Market

respondent might require a unique product space to account for aspects of his perceptions. However, one of the main reasons for product spaces' utility is that they summarize a large amount of information in unusually tangible and compact form. Allowing a totally different product space for each respondent would certainly destroy much of the illustrative value of the result. A compromise would be to recognize that respondents might fall naturally into a relatively small number of subgroups with different product perceptions. In this case, a separate product space could be constructed for each subgroup.

Frequently a single product space is assumed to be adequate to account for important aspects of all respondents' *perceptions*. Differences in *preference* are then taken into account by considering each respondent's idea product to have a unique location in the common product space, and by recognizing that different respondents may weight dimensions uniquely. This was the approach taken in the examples to follow.

Techniques which have received a great deal of use in constructing product spaces include nonmetric multidimensional scaling [3, 7, 8, 12], factor analysis [11], and multiple discriminant analysis [4]. Factor analysis has been available for this purpose for many years, and multidimensional scaling was discussed as early as 1938 [13]. *Nonmetric* multidimensional scaling, a comparatively recent development, has achieved great popularity because of the invention of ingenious computing methods requiring only the most minimal assumptions regarding the nature of the data. Discriminant analysis requires assumptions about the metric properties of data, but it appears to be particularly robust and foolproof in application.

These techniques produce similar results in most practical applications. The technique of multiple discriminant analysis will be illustrated here.

EXAMPLES OF PRODUCT SPACES

Imagine settling on a number of attributes which together account for all of the important ways in which products in a set are seen to differ from each other. Suppose that each product has been rated on each attribute by several people, although each person has not necessarily described more than one product.

Given such data, multiple discriminant analysis is a powerful technique for constructing a spatial model of the product category. First, it finds the weighted combination of attributes which discriminates most among products, maximizing an F-ratio of between-product to within-product variance. Then second and subsequent weighted combinations are found which discriminate maximally among products, within the constraint that they all be uncorrelated with one another. Having determined as many discriminating dimensions as possible, average scores can be used to plot products on each dimension. Distances between pairs of products in this space reflect the amount of discrimination between them.[1]

Figure 1 shows such a space for the Chicago beer market as perceived by members of Market Facts' Consumer Mail Panels in a pilot study, September 1968. Approximately 500 male beer drinkers described 8 brands of beer on each of 35 attributes. The data indicated that a third sizable dimension also existed, but the two dimensions pictured here account for approximately 90% of discrimination among images of these 8 products.

The location of each brand is indicated on these two major dimensions. The horizontal dimension contrasts premium quality on the right with popular price on the left. The vertical dimension reflects relative lightness. In addition, the mean rating of each product on each

---

[1] McKeon [10] has shown that multiple discriminant analysis produces the same results as classic (metric) multidimensional scaling of Mahalanobis' distances based on the same data.

of the atttributes is shown by relative position on each attribute vector. For instance, Miller is perceived as being most popular with women, followed by Budweiser, Schlitz, Hamms, and four unnamed, popularly priced beers.

As a second example, the same technique was applied to political data. During the weeks immediately preceding the 1968 presidential election, a questionnaire was sent to 1,000 Consumer Mail Panels households. Respondents were asked to agree or disagree with each of 35 political statements on a four-point scale. Topics were Vietnam, law and order, welfare, and other issues felt to be germane to current politics. Respondents also described two preselected political figures, according to their perceptions of each figure's stand on each issue. Discriminant analysis indicated two major dimensions accounting for 86% of the discrimination among 14 political figures.

The liberal vs. conservative dimension is apparent in the data, as shown in Figure 2. The remaining dimension apparently reflects perceived favorability of attitude toward government involvement in domestic and international matters. As in the beer space, it is only necessary to erect perpendiculars to each vector to observe each political figure's relative position on each of the 35 issues. Additional details are in [5].

Multiple discriminant analysis is a major competitor of nonmetric multidimensional scaling in constructing product spaces. The principal assumptions which the former requires are that: (1) perceptions be homogeneous across respondents, (2) attribute data be scaled at the interval level (equal intervals on rating scales), (3) attributes be linearly related to one another, and (4) amount of diagreement (error covariance matrix) be the same for each product.

Only the first of these assumptions is required by most nonmetric methods, and some even relax that assumption. However, the space provided by multiple discriminant analysis has the following useful properties:

1. Given customary assumptions of multivariate normality, there is a test of significance for distance (dissimilarity) between any two products.
2. Unlike nonmetric procedures, distances estimated among a collection of products do not depend upon whether or not additional products are included in the analysis. Any of the brands of beer or political figures could have been deleted from the examples and the remaining object locations would have had the same relationships to one another and to the attribute vectors.

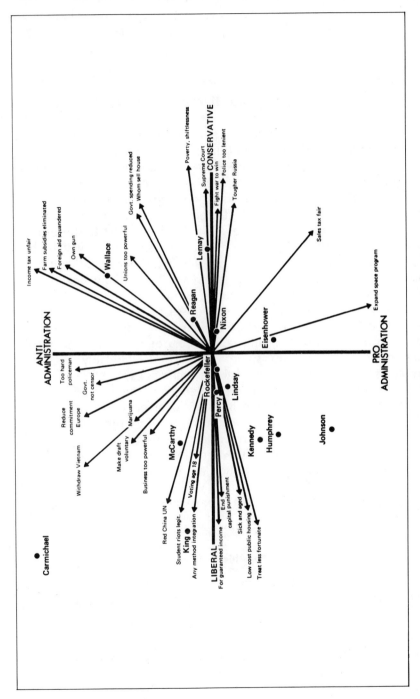

Figure 2—The Political Space, 1968

3. The technique is reliable and well known, and solutions are unique, since the technique cannot be misled by any local optimum.

OBTAINING THE DISTRIBUTION OF CONSUMERS' IDEAL POINTS

After constructing a product space, the next concern is estimating consumer demand for a product located at any particular point. The demand function over such a space is desired and can be approximated by one of several general approaches.

The first to locate each person's ideal point in the region of the space implied by his rank ordered preferences. His ideal point would be closest to the product he likes best, second closest to the product he likes second best, etc. There are several procedures which show promise using this approach [2, 3, 7, 8, 12], although difficulties remain in practical execution. This approach has trouble dealing with individuals who behave in a manner contrary to the basic assumptions of the model, as when one chooses products first on the far left side of the space, second on the far right side, and third in the center. Most individuals giving rank orders of preference do display such nonmonotonicity to some extent, understandably producing problems for the application of these techniques.

The second approach involves deducing the number of ideal points at each region in space by using data on whether a product has too much or too little of each attribute. This procedure has not yet been fully explored, but at present seems to be appropriate to the multidimensional case only when strong assumptions about the shape of the ideal point distribution are given.

The third approach is to have each person describe his ideal product, with the same attributes and rating scales as for existing

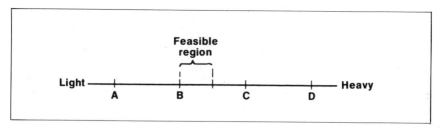

Figure 3—A One-Dimensional Product Space

products. If multiple discriminant analysis has been used to obtain a product space, each person's ideal product can then be inserted in the same space.

There are considerable differences between an ideal point location inferred from a rank order of preference and one obtained directly from an attribute rating. To clarify matters, consider a single dimension, heaviness vs. lightness in beer. If a previous mapping has shown that Brands A, B, C, and D are equally spaced on this one dimension, and if a respondent ranks his preferences as B, C, A, and D, then his ideal must lie closer to B than to A or C and closer to C than to A. This narrows the feasible region for his ideal point down to the area indicated in Figure 3. Had he stated a preference for A, with D second, there would be no logically corresponding position for his ideal point in the space.

However, suppose these products have already been given the following scale positions on a heavy/light dimension: A = 1.0, B = 2.0, C = 3.0, and D = 4.0. If a respondent unambiguously specifies his ideal on this scale at 2.25, his ideal can be put directly on the scale, with no complexities. Of course, it does not follow *necessarily* that his stated rank order of preference will be predictable from the location of his ideal point.

There is no logical reason why individuals must be clustered into market segments. Mathematically, one can cope with the case where hundreds or thousands of individual ideal points are each located in the space. However, it is much easier to approximate such distributions by clustering respondents into groups. Cluster analysis [6] has been used with the present data to put individuals into a few groups with relatively similar product desires (beer) or points of view (politics).

Figure 4 shows an approximation to the density distribution of consumer's ideal points in the Chicago beer market, a "poor man's contour map." Ideal points tended somewhat to group themselves (circles) into clusters. It is not implied that all ideal points lie within the circles, since they are really distributed to some extent throughout the entire space. Circle sizes indicate the relative sizes of clusters, and the center of each is located at the center of the circle.

A representation such as this contains much potentially useful marketing information. For instance, if people can be assumed to prefer products closer to their ideal points, there may be a ready market for a new brand on the lower or "heavy" side of the space, approximately neutral in price/quality. Likewise, there may be op-

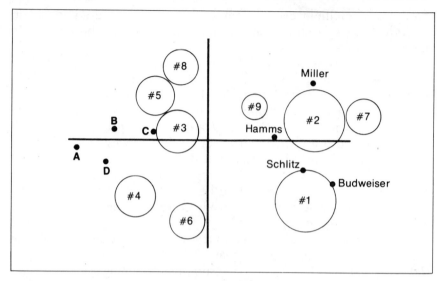

Figure 4—Distribution of Ideal Points in Product Space

portunities for new brands in the upper middle region, decidedly light and neutral in price/quality. Perhaps popularly priced Brand A will have marketing problems, since this brand is closest to no cluster.

Figure 5 shows a similar representation for the political space, where circles represent concentrations of voters' points. These are not ideal points, but rather personally held positions on political issues. Clusters on the left side of the space intended to vote mostly for Humphrey and those on the right for Nixon in the 1968 election. Throughout the space, the percentage voting Republican increases generally from left to right.

It may be surprising that the center of the ideal points lies considerably to the right of that of the political figures. One possible explanation is that this study dealt solely with positions on *issues,* so matters of style or personality did not enter the definition of the space. It is entirely possible that members of clusters one and eight, the most liberal, found Nixon's position on issues approximately as attractive as Humphrey's, but they voted for Humphrey on the basis of preference for style, personality, or political party. Likewise, members of cluster two might have voted strongly for Wallace, given his position, but he received only 14% of this cluster's vote. He may have been rejected on the basis of other qualities. The clusters are described in more detail in [5].

A small experiment was undertaken to test the validity of this

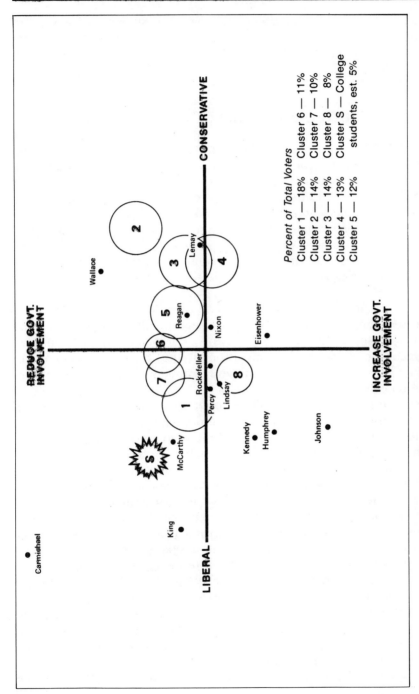

Figure 5—Voter Segment Positions Relative to Political Figures

model. Responses from a class of sociology students in a western state university showed them to be more liberal and more for decreasing government involvement internationally than any of the eight voter clusters. Their position is close to McCarthy's, indicated by an "S."

STRATEGIC INTEGRATION OF FINDINGS

Having determined the position of products in a space and seen where consumer ideal points are located, how can such findings be integrated to determine appropriate product strategy? A product's market share should be increased by repositioning: (1) closer to ideal points of sizable segments of the market, (2) farther from other products with which it must compete, and (3) on dimensions weighted heavily in consumers' preferences. Even those broad guidelines provide some basis for marketing strategy. For instance, in Figure 4, Brand A is clearly farthest from all clusters and should be repositioned.

In Figure 5, Humphrey, Kennedy, and Johnson could have increased their acceptance with this respondent sample by moving upwards and to the right, modifying their perceived position. Presumably, endorsement of any issue in the lower left quadrant of Figure 2 would have helped move Humphrey closer to the concentration of voters' ideal points.

Although the broad outlines of marketing strategy are suggested by spaces such as these, it would be desirable to make more precise quantitative forecasts of the effect of modifying a product's position. Unfortunately, the problem of constructing a model to explain product choice behavior based on locations of ideal points and products in a multidimensional space has not yet been completely solved, although some useful approaches are currently available.

As the first step, it is useful to concentrate on the behavior of clusters of respondents rather than that of individuals, especially if clusters are truly homogeneous. Data predicting behavior of groups are much smoother and results for a few groups are far more communicable to marketing management than findings stated in terms of large numbers of individual respondents.

If preference data are available for a collection of products, one can analyze the extent to which respondents' preferences are related to distances in the space. Using regression analysis, one can estimate a set of importance weights for each cluster or, if desired, for each respondent, to be applied to the dimensions of the product space.

Weights would be chosen providing the best explanation of cluster or individual respondent preferences in terms of weighted distances between ideal points and each product's perceived location. If clusters, rather than individuals, are used, it may be desirable to first calculate preference scale values or utilities for each cluster [1, 9]. Importance weights can then be obtained using multiple regression to predict these values from distances. If explanations of product preference can be made for *existing products,* which depend only on locations in space, then the same approach should permit *predictions* of preference levels for new or modified products to be positioned at specific locations in the space.

Models of choice behavior clearly deserve more attention. Although the problem of constructing the product space has received much attention, we are denied the full potential of these powerful solutions unless we are able to quantify relationships between distances in such a space and consumer choice behavior.

SUMMARY

Market segmentation studies can produce results which indicate desirable marketing action. Techniques which are presently available can: (1) construct a product space, (2) discover the shape of the distribution of consumers' ideal points throughout such a space, and (3) identify likely opportunities for new or modified products.

In the past, marketing research has often been restricted to *tactical* questions such as package design or pricing levels. However, with the advent of new techniques, marketing research can contribute directly to the development of *strategic* alternatives to current product marketing plans. There remains a need for improved technology, particularly in the development of models for explaining and predicting preferential choice behavior. The general problem has great practical significance, and provides a wealth of opportunity for development of new techniques and models.

REFERENCES

1. Bradley, M. E. and R. A. Terry. "Rank Analysis of Incomplete Block Designs: The Method of Paired Comparisons," *Biometrika,* 39 (1952), 324–45.
2. Carroll, J. D. "Individual Differences and Multidimensional Scaling," Murray Hill, N.J.: Bell Telephone Laboratories, 1969.
3. Guttman, Louis. "A General Nonmetric Technique for Finding the Smallest Space for a Configuration of Points," *Psychometrika,* 33 (December 1968), 469–506.

4. Johnson, Richard M. "Multiple Discriminant Analysis," unpublished paper, Workshop on Multivariate Methods in Marketing, University of Chicago, 1970.
5. ———. "Political Segmentation," paper presented at Spring Conference on Research Methodology, American Marketing Association, New York, 1969.
6. Johnson, Stephen C. "Hierarchial Clustering Schemes," *Psychometrika*, 32 (September 1967), 241–54.
7. Kruskal, Joseph B. "Multidimensional Scaling by Optimizing Goodness of Fit to a Nonmetric Hypothesis," *Psychometrika*, 29 (March 1964), 1–27.
8. ———. "Nonmetric Multidimensional Scaling: A Numerical Method," *Psychometrika*, 29 (June 1964), 115–29.
9. Luce, R. D. "A Choice Theory Analysis of Similarity Judgments," *Psychometrika*, 26 (September 1961), 325–32.
10. McKeon, James J. "Canonical Analysis," *Psychometric Monographs*, 13.
11. Tucker, Ledyard. "Dimensions of Preference," Research Memorandum RM-60-7, Princeton, N.J.: Educational Testing Service, 1960.
12. Young, F. W. "TORSCA, An IBM Program for Nonmetric Multidimensional Scaling," *Journal of Marketing Research*, 5 (August 1968), 319–21.
13. Young, G. and A. S. Householder. "Discussion of a Set of Points in Terms of Their Mutual Distances," *Psychometrika*, 3 (March 1938), 19–22.

CONJOINT MEASUREMENT FOR QUANTIFYING JUDGMENTAL DATA

*PAUL E. GREEN and VITHALA R. RAO**

The quantification of managerial or consumer judgment has long posed problems for marketing researchers, irrespective of their interest in normative or descriptive decision making. For example, most media selection models reflect some dependence on media planners' judgmental estimates [4, 5], which are often used in evaluating target populations in terms of households' or individuals' product usage or demographic and socio-economic characteristics [6]. Moreover, subjective weights are frequently used in appraising vehicle appropriateness and advertising message perception.

* Paul E. Green is Professor of Marketing, Wharton School of Finance and Commerce, University of Pennsylvania. Vithala R. Rao is Assistant Professor of Marketing, Cornell University. They are indebted to the American Association of Advertising Agencies' Educational Foundation and the General Electric Foundation for providing partial financial support for this project.
Reprinted by permission of the authors and publisher from the Journal of Marketing Research, *published by the American Marketing Association Vol. viii (August 1971) pp. 355–63.*

Further, the study of consumer decision making requires ascertaining how buyers trade off conflicting criteria in making purchase decisions [8]. Finally, recent studies in public administration attest to a growing interest in the modeling of administrators' evaluations involving multiattribute alternatives in which the analyst must rely on judgmental estimates to a large extent [10].

The purpose of this article is to describe a new approach to quantifying judgmental data, conjoint measurement. Its procedures require only rank-ordered input, yet yield interval-scaled output.[1] The principles of conjoint measurement are discussed and synthetic data are used in solving some typical problems. The conclusion is a discussion of limitations of these techniques and potential areas of application to marketing planning and other types of choice behavior.

CONJOINT MEASUREMENT

As the name suggests, conjoint measurement is concerned with the joint effect of two or more independent variables on the *ordering* of a dependent variable. For example, one's preference for various houses may depend on the joint influence of such variables as nearness to work, tax rates, quality of school system, anticipated resale value, and so on. Starting with the theoretical work of Luce and Tukey [14], mathematical psychologists have developed procedures for simultaneously measuring the joint effects of two or more variables at the level of interval scales (with common unit) from rank-ordered data alone.

An important special case of conjoint measurement is the *additive* model, which is analogous to the absence of interaction in the analysis of variance involving two (or more) levels of two (or more) factors in a completely crossed design [2]. In the latter procedure one tests whether or not original cell values can be portrayed as additive combinations of row and column effects. In additive conjoint measurements, however, one asks if the cell values can be monotonically transformed so that additivity can be achieved.[2]

Since the work of Luce and Tukey, mathematical psychologists

---

[1] In the case of finite data, the scale is technically an ordered metric; as the number of input values increases, however, a unique representation at the interval scale level is approached.

[2] The typical handling of judgmental estimates in the media models examined here entails developing numerical estimates on a single-factor-at-a-time basis, i.e., *without* explicit consideration of interactions. In this regard the additivity assumption described here does not appear unwarranted.

TABLE 1
RANK ORDER INPUT DATA FOR PROBLEM 1

Ads	Vehicles				
	1	2	3	4	5
1	1[a]	3	8.5	16.5	19.5
2	2	5	11	21	24.5
3	4	7	13	24.5	26
4	6	10	15	28	30
5	8.5	12	19.5	31.5	33
6	14	18	28	35	36
7	16.5	22.5	31.5	37	38
8	22.5	28	34	39	40

[a] Rank 1 indicates least effective ad-vehicle combination.

have extended additive conjoint models to deal with nonadditivity, partially ordered data, and any polynomial type of function. Analogous to our discussion of the additive model, a data matrix satisfies the (more general) polynomial [3] model whenever it is possible to rescale each cell entry so that it is represented by a specified polynomial function of the row and column variables, and the representation preserves the rank order of the original cell entries as closely as possible [23, 26].

Without diminishing the importance of these extensions, the additive case seems to have proven quite useful in a variety of applications [3, 11, 18, 22]; thus in subsequent discussion we emphasize this special case.

Algorithms for Conjoint Measurement

The work of Luce and Tukey can be appropriately characterized as providing the conceptual foundations of conjoint measurement. One still requires algorithmic procedures for finding *numerical* representations that best satisfy the conditions described above. Recently a variety of algorithms have been developed by many of the same researchers [12, 13, 16, 24, 26] who have contributed to the (closely allied) field of nonmetric multidimensional scaling.

Kruskal's MONANOVA algorithm was used to illustrate the solu-

---

[3] A "polynomial" function involves a specific combination of sums, differences, and products of its arguments.

TABLE 2
"ORIGINAL" SCALE VALUES FOR PROBLEM 1

Ad impact values	Vehicle appropriateness values				
	$b_1 = 2$	$b_2 = 6$	$b_3 = 13$	$b_4 = 22$	$b_5 = 24$
$a_1 = 1$	3	7	14	23	25
$a_2 = 4$	6	10	17	26	28
$a_3 = 6$	8	12	19	28	30
$a_4 = 9$	11	15	22	31	33
$a_5 = 12$	14	18	25	34	36
$a_6 = 18$	20	24	31	40	42
$a_7 = 21$	23	27	34	43	45
$a_8 = 25$	27	31	38	47	49

tion of Problems 1 and 2 (to follow) [12]. Problem 3 can be solved by a variety of nonmetric scaling algorithms. One such algorithm, Young and Torgerson's TORSCA [27] program, is employed illustratively here.

APPLYING CONJOINT MEASUREMENT PROCEDURES

Problem 1

A media planner must select medical journals for promoting a new ethical drug. Specifically, he is interested in assigning importance values to 8 print advertisements (whose appeals vary) and 5 journals that represent possible vehicles for the candidate ads. He feels that he can rank the 40 ad-vehicle combinations in terms of overall effectiveness, but is not confident about assigning numerical (interval-scaled) weights to ad impact [4] or vehicle appropriateness, either on a one-variable-at-a-time basis or in combination. However, he would ultimately like to have numerical values (interval scaled and expressed in terms of a common unit) of *both* ad impact and vehicle appropriateness.

To make the discussion more transparent, the values in Table I are assumed to represent the media planner's ranking of the 40 ad-vehicle combinations in terms of overall effectiveness. In actuality the rank numbers of Table 1 were obtained by merely transforming the

---

[4] Ad impact is defined as the relative value of an ad in prompting perception of the message. Presumably ad impact reflects the judger's evaluation of both its thematic and physical aspects [19].

cell entries of Table 2, which *do* represent additive combinations of row and column effects, to integer ranks (including ties). If the procedure works, one would expect to "recover" the original row and column scales up to positive linear transformations (with common unit but arbitrary origins).

Kruskal's MONANOVA algorithm was applied to the ranked data of Table 1. The stress (badness of fit) was virtually zero (actually 0.002), as would be expected given the error-free data used in this example.

The upper panel of Figure 1 shows scatter plots of the main effects (estimated by the program) on the original row and column values, respectively, of Table 2. Also shown (in the lower panel of Figure 1) is the monotone function that relates the input data from Table 1 to the cell values that are estimated by the model. This function is also found by the program.

At this point, then, the media planner has developed:

1. Separate interval scales for ad impact (the row values) and vehicle appropriateness (the column values), as shown by projections onto the vertical axes of the two scatter diagrams in Figure 1. Moreover, these scales are expressed in terms of a common unit.

2. Interval-scaled cell entries for each of the 40 ad-vehicle combinations. These represent the simple sum of the derived row and column scales whose values are shown on the vertical axes in the two scatter diagrams.

The above interval-scaled values can then be used as direct numerical judgments, which are used in various media selection models.

At this point it is useful to discuss the motivation and assumption structure underlying the application of additive conjoint measurement. First, we assume that the media planner would find direct numerical rating of the ads and vehicles too difficult; second, if he merely ranked ads and vehicles separately, he would *not* have enough information to derive interval scales. Ranking conjoint pairs (ad-vehicle combinations) provides sufficient information for finding separate *numerical* scales for each.

The additive conjoint measurement model entails the following assumptions:

1. The data matrix (Table 1) contains elements that can be at least weakly ordered (may contain ties).

2. Each derived cell value represents an additive combination of separate (real-valued) functions of the row and column elements.

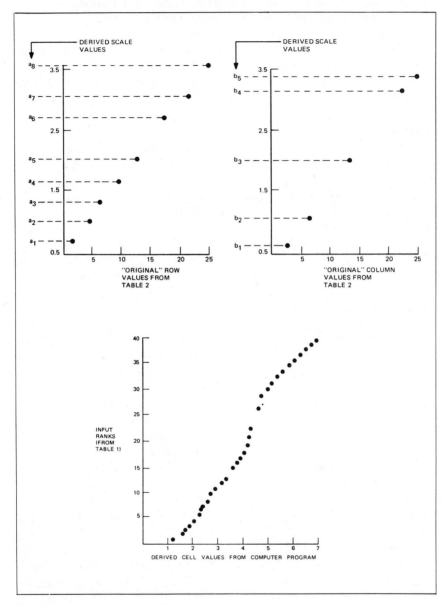

Figure 1—Scatter Plots of Derived vs. Original Row and Column Effects and Input Ranks vs. Derived Cell Values

3. Each numerical cell value in the derived solution comes as close as possible to maintaining the rank order of the input data.[5]

Thus the additive conjoint model is a monotone analogue of main-effects analysis of variance. In Problem 1 the arguments of the row and column functions are nominal variables, viz. the specific ads and vehicles.

Problem 2

A marketing researcher is interested in finding component utilities (or part-worths) that housewives attribute to various characteristics of discount cards, e.g., size of discount, number of cooperating stores in the trading area, and initial cost of the card. Sample discount cards are prepared whose levels on each of the above characteristics are systematically varied on the basis of four levels for the first factor and three levels each for the second and third factors. A housewife is asked to rank the resulting 36 cards in terms of "best buy for the money." The researcher is interested in deriving the housewife's component utilities (positive or negative) for each of the three factors considered jointly.

This problem differs structurally from the previous one. First, it is represented by a three-way ($4 \times 3 \times 3$) rather than two-way design. Second, the arguments of the contributory factors—size of discount, number of cooperating stores, and initial cost of card—are ratio rather than nominal-scaled variables.

Table 3 illustrates the rank order data that could be gathered in this type of situation, and Figure 2 shows how these were actually generated. For each factor we assume that the housewife has an implicit component utility (or disutility) function, shown by the curved lines in Figure 2. While she cannot explicate these functions, her over-all ranking of the discount cards reflects an additive combination of the values of each individual utility function for the discrete levels of each factor utilized in the experiment. For example, the first cell value of 34 in the lower panel of Figure 2 is the sum of 11, 4 and 19, the component utilities shown in the upper panels. The values of these functions are indicated by dotted lines. Note that the functions need not be linear or even monotone (as illustrated for the size-of-discount factor).

---

[5] Necessary and sufficient conditions for the existence of such additive transforms can be found in [2]. The axiomatic structure for the more general polynomial model appears in [23].

In this case we wished to find the selected values of each component utility function for selected levels of each factor. The ranked data of Table 3 were again processed [12] and the results are shown in Figure 3. The upper panel shows "recovered" component utilities. Note that the derived utilities match closely (up to a linear transformation with common unit) the original values shown in Figure 2. The lower panel of Figure 3 shows the appropriate monotone transform that links the derived cell values to the original ranked data of Table 3.

Problem 2 illustrates that additive conjoint measurement provides a potentially useful tool for estimating component utilities from preference for a composite (complete) discount card profile.

Problem 3

A researcher in public administration is interested in developing an interval scale of law enforcement officers' opinions of the seriousness of various forms of drug abuse, e.g., marijuana, amphetamines, hashish, heroin, etc. Eight such drugs have been listed; for each pair of drugs the respondent is asked to state: (1) which drug of each pair is more serious (in terms of harm to the user) and (2) the intensity of difference in seriousness, expressed as a rank number on an ordered-category scale. From such ranked values the researcher would like to develop a unidimensional interval scale on which the eight drugs can be positioned in terms of perceived personal harm.

Problem 3 illustrates the more general (polynomial) form of conjoint measurement. Table 4 shows the law enforcement officers' set of responses of the intensity with which each pair of drugs is separated in terms of seriousness. For example, suppose he judged Drug 2

TABLE 3
RANK ORDER INPUT DATA FOR PROBLEM 2

Size of discount	Cost of card								
	5 stores			10 stores			15 stores		
	$7	$14	$21	$7	$14	$21	$7	$14	$21
5%	9.5	5	1[a]	18	12	3.5	29.5	24	12
10%	18	12	3.5	27	21.5	8	34.5	31.5	21.5
20%	24	18	6.5	31.5	27	14	36	34.5	27
30%	15	9.5	2	24	18	6.5	33	29.5	18

[a] Rank 1 indicates worst buy for the money.

to be more serious than Drug 1. In addition, he assigned a rank of 1 on a 7-point intensity scale to reflect his feeling that the difference in seriousness is quite low in intensity. Thus the first cell shows the value of 1.

As before, Table 4 was derived from the synthetic data of Table 5, in which the (assumed) interdrug distances are shown. For example, the first cell value of 3 in Table 5 is the distance between Drug 1 and Drug 2 (3 = 4 − 1). Note that the ranked data of Table 4 do not tell us, per se, what the rank order of values (on a scale of seriousness) is, as derived from the respondent's paired comparisons. However, in the errorless case described here this rank order will be preserved.

The data matrix of Table 4 was then submitted to the TORSCA program to find a one-dimensional solution [27]. The stress value was 0.082, reflecting the large number of ties in the input data.

Despite the somewhat high stress value, recovery of the unidimensional scale—both in rank order and scale values—is excellent.

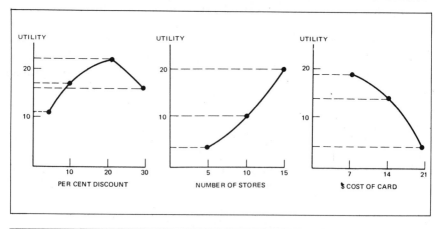

Size of	Cost of card								
discount	5 stores			10 stores			15 stores		
	$7	$14	$21	$7	$14	$21	$7	$14	$21
5%	34	29	19	40	35	25	50	45	35
10%	40	35	25	46	41	31	56	51	41
20%	45	40	30	51	46	36	61	56	46
30%	39	34	24	45	40	30	55	50	40

Figure 2—Generation of Original Values for Problem 2 and Cell Values Obtained from Component Utilities

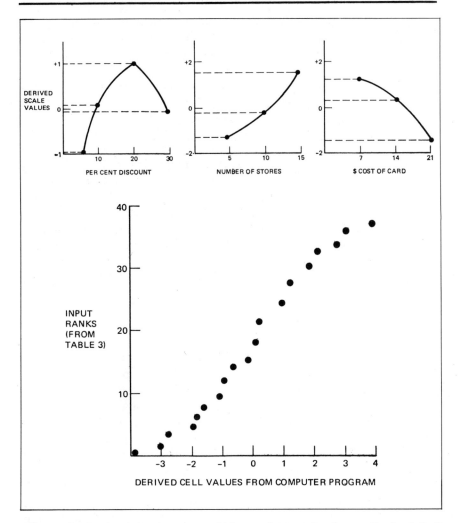

.**Figure 3**—Derived Component Utilities and Input Ranks vs. Derived Cell Values

The left-hand panel of Figure 4 shows a scatter plot of the derived scale values (from TORSCA) vs. the original scale values. The linearity between the two scales is quite evident. As a matter of interest, the product moment correlation between derived and original interpoint *distances* is 0.99. The right-hand panel shows a plot of input data ranks (shown originally in Table 4) on interpoint distances obtained from the scaling program, showing the specific monotone transform that links each set of values.

Problem 3, though viewed as a scaling-type problem, illustrates a more general combinatorial rule than the additive rule utilized in solving Problems 1 and 2. In this case the joint effect (as noted in Table 4) is in terms of a monotone transform function that is represented by the square root of the squared difference of each pair of scale values on a single dimension,[6] (the original scale of Table 5). In this sense nonmetric scaling, undimensional or multidimensional, can be viewed as a special case of polynomial conjoint measurement.

LIMITATIONS OF THE TECHNIQUES

As might have already been surmised, the additive assumption used in solving Problems 1 and 2 may be unduly restrictive: bona fide interactions may be present. If so, more general polynomial models are called for.[7] However, we suspect that in many instances the simpler (additive) model represents a very good approximation of reality. That is, what are often called "interactions" in traditional ANOVA applications may be the result of our failure to measure the effects of independent variables on the correct scales in the first place [9].

However, as suggested by Problem 3, the respondent may not behave unidimensionally toward the prespecified criterion. This could be reflected in the failure of the paired comparisons to generate a complete rank order (necessitating the use of nearest adjoining order techniques [20]) or the failure of a one-dimensional scaling solution to accommodate the data. If such results occur, it seems to us that the researcher should recognize the multidimensional nature of the criterion and deal with it as such (e.g., as a linear combination of more basic dimensions). Some of our recent research suggests that multidimensional scaling results *can* occur in the context of so-called unidimensional constructs that are often used for rating scale purposes.

Third, in dealing with large-size problems (Problem 1 or Problem 2) the ranking task becomes formidable. Here we would favor the use of ordered categories, say, on a 9 or 11-point scale, in which each combination of ad-vehicles would receive a rating of favorableness. Moreover, some comparisons could be omitted, inasmuch as the programs tolerate missing data. However, relatively little is known about which (and how many) comparisons are the best ones to omit [8]. Finally, one could use various designs—fractional factorials—that

---

[6] In the one-dimensional case all Minkowski metrics yield the same result.

[7] Recent developments [26] have provided appropriate algorithms for the more general case in which "genuine" interactions may exist.

reduce the number of combinations while still permitting the "esti-mation" of main effects.

Fourth, as in nonmetric scaling, conjoint solutions are susceptible to certain types of degeneracy; some of these cases are described by Kruskal [12].

Finally, it is still a moot point as to whether direct numerical estimation procedures would lead to results comparable to those arrived at through ranking followed by conjoint measurement. We have assumed throughout the exposition that it is easier for a judge to rank items than it is to provide direct numerical values. At this stage of methodological development it seems that cross-comparison of procedures would be needed before one could make any definitive evaluation of the merits (or demerits) of conjoint measurement procedures vs. direct numerical estimation.

OTHER AREAS OF APPLICATION

Before commenting on other areas of potential application, it should be mentioned that the approach illustrated above can be generalized to deal with: (1) more than three independent variables and (2) more than a single assessor. An illustration of the first case is where the discount card profile is further characterized by: (1) nearness of stores to respondent's home and (2) type of cooperating store, e.g., specialty stores only vs. specialty and department stores. No new principles are involved; in this case a five-way (rather than three-way) matrix would be set up for evaluation. (The implied ranking task might be quite formidable, however, unless some type of fractional factorial design were used.)

If several assessors were involved one could: (1) develop each evaluator's scale values separately, (2) sum the rankings over assessors and do a group analysis, or (3) use a preliminary cluster analysis and develop a limited number of points of view with a separate set of scales for each. We favor the third procedure, particularly because of its value in linking between-group differences with other (background) characteristics of the evaluators.

It is not at all difficult to imagine other areas where conjoint measurement could be used in marketing research. The following list is meant to illustrate potential content areas and is included largely to stimulate the reader's curiosity about the applicability of the methodology to his own areas of interest. We emphasize the fact that the proposed applications are speculative—empirical research in conjoint measurement is just beginning.

Vendor Evaluations

In industrial marketing, conjoint measurement could be used in developing purchasing agents' ratings of (hypothetical) vendors on a variety of evaluative scales, such as delivery reliability, product quality, technical service back-up, and so on. For example, this could be done by setting up vendor profiles of, say, five factors, each at two levels of effectiveness, a total of 2^5 or 32 combinations [25]. By using a procedure similar to that in Problem 2, one could obtain component utilities for each level of each of the factors. These, in turn, could be utilized in developing effectiveness measures for real vendors.

Price-Value Relationships

A combination of additive conjoint measurement and intensity scaling could be used to measure consumers' evaluations of price-value [17]. First, intensity scaling (as applied in Problem 3) could be used to develop a unidimensional scale of quality for, say, a set of brands of electric dryers without any (explicit) price information about the brands. Then the respondent would be shown combinations of the brands and prices (e.g., all crossed combinations of 6 dryers and 6 price levels) and asked to rank the 36 brand-price combinations with regard to best value for the money. In this way one could obtain conjoint scales of price disutility and quality utility, as well as the psychophysical transforms that relate each, respectively, to objective price and the previously obtained quality scale.

Bayesian Prior Estimation

Intensity scaling could be used to develop Bayesian prior distributions by generalizing Smith's procedure to a higher-ordered metric [7, 15, 21]. The resulting interval scale would require some direct estimation to establish an origin in order to convert the interval-scaled values to a ratio scale. Moreover, by using more general conjoint measurement models (e.g., multiplicative), other types of probability assessment could be made as well (e.g., joint probabilities). Alternatively, log transformations could be used [23].

New Venture Appraisal

Conjoint measurement techniques could also be used to measure a decision maker's trade-offs between mean and variance of cash flow by showing him an array of cash flow distributions varying in mean and dispersion (or, if desired, additional moments of the distribution). The additive conjoint model could then be applied to assess the

TABLE 4
RANK ORDER INPUT DATA FOR PROBLEM 3

Drugs	Drugs						
	2	3	4	5	6	7	8
1	1[a]	2	4	4	6	7	7
2		1[a]	2	3	5	6	7
3			1[a]	3	5	6	7
4				1[a]	4	4	6
5					3	4	5
6						2	3
7							2

[a] Rank 1 stands for very little difference in seriousness between members of a pair of drugs.

component utilities of each level of each moment of the probability distribution in overall preference.

In a variant of this procedure, the assessor would rank research project descriptions and cost outlay combinations in terms of likelihood of successful completion over a target planning period. Or his preference for alternative "futures" characterized by scenarios containing profile descriptions of company position and various environmental variables could be determined. This last application could have relevance for long-range planning studies.

Promotional Congruence Testing

Still another area of application concerns measurement of the joint effects of congruent characteristics—package design, copy theme, price, brand image—on the overall evaluation of a brand. The relationship among congruent (vs. discrepant) elements of the marketing mix could be assessed in terms of the component scale values derived from conjoint measurement models. In principle, advertising campaigns could be designed around those promotional elements that, in combination, produce the highest consumer evaluations for a specified target group and implementation cost.

Attitude Measurement

Many approaches to scaling attitudes toward objects assume a type of additive model in which total affect is represented by a linear combination of evaluative beliefs. For example, brands may be described in terms of a set of attributes and respondents asked to rate each brand: (1) with regard to the level of each attribute and (2) in terms of overall value. Multiple regression may then be used to solve for "importance weights" (the regression coefficients) for assigning to each attribute in order to maximize the correlation between overall worth and a linear combination of the attribute ratings.

This type of problem can be handled within the conjoint measurement framework by first using the scaling approach associated with Problem 3 to obtain attribute scale values. Then one can use the approach associated with Problem 2 to find part-worths for each evaluative belief, utilizing, for example, additive conjoint measurement. As such, the subject's overall evaluations need only be rank ordered, i.e., monotone regression would replace linear regression under this approach.

Functional vs. Symbolic Product Characteristics

One of the perennial problems in marketing concerns the relationship betwen functional vs. symbolic characteristics of brands in consumer evaluation. In principle, this problem could be approached by first characterizing each (unidentified) brand in terms of a functional, or performance, profile. Brand "quality" scores could then be developed by obtaining consumer responses to various profile-price combinations according to a factorial design, similar to those already discussed.

Then, in a second phase of the experiment, the same subjects would be shown brand-price combinations and a new set of brand "quality" scores obtained for each (now identified) brand. Differences between each brand's score (identified and unidentified) may be taken as a rough approximation of "symbolic" quality. Moreover,

TABLE 5
"ORIGINAL" DISTANCE VALUES FOR PROBLEM 3

Scale value of drugs	d_2	d_3	d_4	Drugs d_5	d_6	d_7	d_8
$d_1 = 1$	3	5	8	11	17	20	24
$d_2 = 4$		2	5	8	14	17	21
$d_3 = 6$			3	6	12	15	19
$d_4 = 9$				3	9	12	16
$d_5 = 12$					6	9	13
$d_6 = 18$						3	7
$d_7 = 21$							4
$d_8 = 25$							0

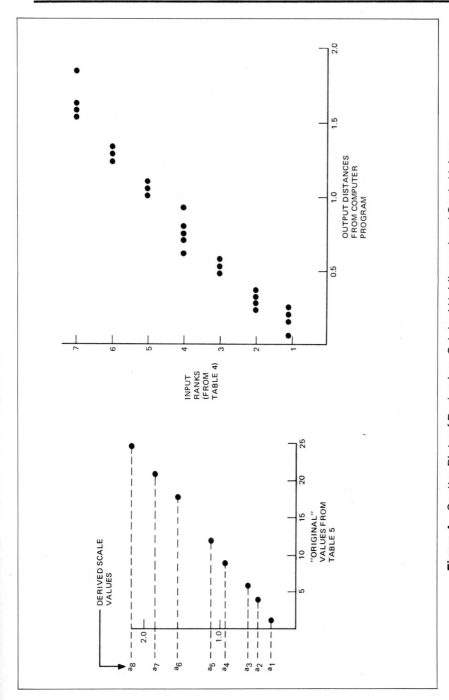

Figure 4—Scatter Plots of Derived vs. Original Unidimensional Scale Values and Input Ranks vs. Output Distances

each subject could be characterized in terms of the extent to which functional vs. symbolic characteristics are important to her overall evaluation. Finally, the approach might be extended across product classes in an attempt to develop indices of brand differentiation at the product class level.

Utility for Item Collections

Conjoint measurement could also be used as a complement to multidimensional scaling procedures in measuring the utility of collections of items. For example, multidimensional scaling may be used to scale separate sets of items, e.g., appetizers, entrees, and desserts. As might be surmised, the perceptual dimensions would probably differ across the three sets. Conjoint measurement could then be used to find utility scales for items of each set in the context of a complete menu of appetizer, entree, and dessert. These scales could then be related to the separate spaces obtained from the multidimensional scaling procedures.

If all items are from the same set, e.g., cereals, conjoint methods could still be used in conjunction with multidimensional scaling to develop utility scales in the context of collections of items—ideal assortments of cereals for combination packaging. It is possible that this approach may yield insight into such interesting problems as the utility for variety and the relationship of assortment values to component values when the latter are considered as alternative first choices (the usual assumption in the scaling of preference data).

Cost-Benefit Analysis

To us one of the most exciting areas of potential application is the use of conjoint measurement models in assessing trade-off relationships in the development of multiattribute effectiveness measures. As an illustration, suppose a number of alternative law enforcement plans were designed and characterized by such attributes as: (1) crime rate, by major category; (2) criminal arrest rate; (3) criminal rehabilitation rate; (4) cost, and so on. Again, in principle, one could develop public officials' trade-off relationships in terms of the implied component utilities developed from applying conjoint measurement models. One study has demonstrated the feasibility of this approach in education [1].

Analysis of Variance

The preceding areas have emphasized behavioral data, but additive conjoint measurement can also be used in analyzing factorial experiments, where the analysis is similar in spirit to ANOVA models.

Perhaps the most useful procedure would be to apply the conjoint measurement model as a preliminary technique in order to see if a monotone transform is sufficient to make the cell entries additive combinations of row and column effects (in the two-way design). Since conjoint measurement algorithms include a display of the best fitting monotone transform, in discrete form, of course, this by-product is useful in the selection of a *specific* functional form— logarithmic, square root, etc.—prior to conducting a standard ANOVA test.

CONCLUSIONS

The primary purpose of this article has been to introduce marketing researchers to some of the concepts and potential applications of conjoint measurement. It seems to us that various types of marketing planning models and other procedures using judgmental estimates in a formal manner might benefit from the utilization of conjoint models—additive or, more generally, polynomial. Moreover, buyer preferences for multiattribute items may also be decomposed into part-worth evaluations in a similar manner.

Too little is known at this time about the relative advantages and disadvantages of these techniques compared to procedures using direct numerical estimation. Hopefully, some of the areas of application mentioned in this article will pique researchers' interests enough to prompt more thorough investigation of some of the research possibilities raised in this overview article.

REFERENCES

1. Carmone, Frank J. "Interaction Budgeting and Studies of the Structure of Subjective Evaluation Functions of Faculty Members," unpublished doctoral dissertation, University of Waterloo, 1971.
2. Coombs, Clyde H., Robyn M. Dawes, and Amos Tversky. *Mathematical Psychology, an Elementary Introduction.* Englewood Cliffs, N.J.: Prentice-Hall, 1970.
3. Coombs, Clyde H. and S. S. Komorita. "Measuring Utility of Money Through Decisions," *American Journal of Psychology,* 71 (August 1958), 383–9.
4. Day, Ralph L. "Linear Programming in Media Selection," *Journal of Advertising Research,* 2 (June 1962), 40–4.
5. Gensch, Dennis H. "Computer Models in Advertising Media Selection," *Journal of Marketing Research,* 5 (November 1968), 414–24.
6. ———. "Media Factors: A Review Article," *Journal of Marketing Research,* 7 (May 1970), 216–25.
7. Green, Paul E. "Critique of Ranking Procedures and Subjective Probability Distributions," *Management Science,* 14 (December 1967), B-250–2.

8.   ———— and Frank J. Carmone, *Multidimensional Scaling and Related Techniques in Marketing Analysis.* Boston: Allyn and Bacon, 1970.
9.   Green, Paul E. and Vithala R. Rao, "Nonmetric Approaches to Multivariate Analysis in Marketing," working paper, University of Pennsylvania, 1969.
10. Huber, George P., Vinod K. Sahney, and David L. Ford. "A Study of Subjective Evaluation Models," *Behavioral Science,* 14 (November 1969), 483–9.
11. Krantz, David H. "Conjoint Measurement: The Luce-Tukey Axiomatization and Some Extensions," *Journal of Mathematical Psychology,* 1 (July 1964), 1–27.
12. Kruskal, Joseph B. "Analysis of Factorial Experiments by Estimating Monotone Transformations of the Data," *Journal of the Royal Statistical Society,* Series B, 27 (March 1965), 251–63.
13. Lingoes, James C. "An IBM-7090 Program for Guttman-Lingoes Conjoint Measurement, I," *Behavioral Science,* 12 (November 1967), 501–2.
14. Luce, R. Duncan and John W. Tukey. "Simultaneous Conjoint Measurement: A New Type of Fundamental Measurement," *Journal of Mathematical Psychology,* 1 (February 1964), 1–27.
15. Morrison, Donald G. "Critique of Ranking Procedures and Subjective Probability Distribution," *Management Science,* 14 (December 1967), B-253–4.
16. Pennell, Roger. "Additive Representations for Two-Dimensional Tables," *Research Bulletin RB-7-29,* Princeton: Educational Testing Service, 1970.
17. Rao, Vithala R. "The Salience of Price in the Perception and Evaluation of Product Quality: A Multidimensional Measurement Model and Experimental Test," unpublished doctoral dissertation, University of Pennsylvania, 1970.
18. Scott, Dana. "Measurement Models and Linear Inequalities," *Journal of Mathematical Psychology,* 1 (July 1964), 233–48.
19. Simon, Leonard S. and Marshall Freimer, *Analytical Marketing.* New York: Harcourt, Brace & World, 1970.
20. Slater, Patrick. "Inconsistencies in a Schedule of Pair Comparisons," *Biometrika,* 48 (December 1961), 303–12.
21. Smith, Lee H. "Ranking Procedures and Subjective Probability Distributions," *Management Science,* 14 (December 1967), B-236–49.
22. Tversky, Amos. "Additivity, Utility and Subjective Probability," *Journal of Mathematical Psychology,* 4 (June 1962), 175–201.
23. ————. "A General Theory of Polynomial Conjoint Measurement," *Journal of Mathematical Psychology,* 4 (February 1967), 1–20.
24. ———— and A. A. Zivian. "A Computer Program for Additivity Analysis," *Behavioral Science,* 11 (January 1966), 78–9.
25. Wind, Yoram, Paul E. Green, and Patrick J. Robinson. "The Determinants of Vendor Selection: The Evaluation Function Approach," *Journal of Purchasing,* 4 (August 1968), 29–41.
26. Young, Forrest W. "Polynomial Conjoint Analysis of Similarities: Definitions for a Specific Algorithm," Research Paper No. 76, Psychometric Laboratory, University of North Carolina, 1969.
27. ———— and Warren S. Torgerson. "TORSCA, A FORTRAN IV Program for Shepard-Kruskal Multidimensional Scaling Analysis," *Behavioral Science,* 12 (November 1967), 498.

BIBLIOGRAPHY

Aaker, D. A. *Multivariate Analysis in Marketing: Theory and Application.* Belmont, Calif.: Wadsworth Publishing Company, 1971.

————. "Visual Clustering Using Principal Component Analysis." In D. A. Aaker (ed.), *Multivariate Methods in Marketing.* Belmont, Calif.: Wadsworth Publishing Company, 1971.

Alpert, M. I. and Peterson, R. A. "On the Interpretation of Canonical Analysis." *Journal of Marketing Research,* 9 (1972), 187-192.

Anderberg, M. R. *Cluster Analysis for Applications.* New York: Academic Press, 1973.

Anderson, T. W. *The Statistical Analysis of Time Series.* New York: Wiley, 1971.

————. "The Use of Factor Analysis in the Statistical Analysis of Multiple Time Series." *Psychometrika,* 28 (1963), 1-25.

————. *An Introduction to Multivariate Statistical Analysis.* New York: Wiley, 1958.

————. "Classification by Multivariate Analysis." *Psychometrika,* 16 (1951), 31-50.

Anderson, T. W. and Rubin, H. "Statistical Inference in Factor Analysis." *Proceedings of the Third Berkeley Symposium on Mathematical Statistics and Probability,* 5 (1956), 111-150.

Andrews, F., Morgan, J. and Sonquist, J. *Multiple Classification Analysis.* Ann Arbor: Institute for Survey Research, University of Michigan, 1969.

Armstrong, J. S. and Soelberg, P. "On the Interpretation of Factor Analysis." *Psychological Bulletin,* 70 (1968), 361-364.

Attneave, F. "Dimensions of Similarity." *American Journal of Psychology,* 63 (1950), 516-556.

Ball, G. H. "Data Analysis in the Social Sciences—What About the Details?" *Proceedings of the Fall Joint Computer Conferences,* Stanford, (1965), 533-559.

Ball, G. H. and Friedman, H. "On the Status of Applications of Clustering Techniques to Behavioral Sciences Data." *Proceedings of the American Statistical Association, Social Statistics,* (1968), 34.

Ball, G. H. and Hall, D. "A Clustering Technique for Summarizing Multivariate Data." *Behavioral Science,* 12 (1967), 153.

Banks, S. "Patterns of Daytime Viewing Behavior," In M. S. Moyer and R. E. Vosburgh (eds.) *Marketing for Tomorrow—Today.* Chicago: American Marketing Association, 1967.

Bartlett, M. S. "The Statistical Significance of Canonical Correlations." *Biometrika,* 32 (1941), 29-38.

Bass, F. M. "A Simultaneous Equation Regression Study of Advertising and Sales of Cigarettes." *Journal of Marketing Research,* 6 (1969), 291-300.

Bass. F. M. and Parsons, L. J. "A Simultaneous-Equation Regression Analysis of Sales and Advertising." *Applied Economics,* 1 (1969), 103-124.

Bass, F. M., Pessemier, E. A., and Tigert, D. J. "A Taxonomy of Magazine Readership Applied to Problems in Marketing Strategy and Media Selection." *Journal of Business,* 42 (1969), 337-363.

Bay, K. S. "Applications of Multivariate Analysis of Variance to Educational and Psychological Research: Hypothesis Testing by MANOVA Programs." *Multivariate Behavioral Research,* 6 (1971), 325-362.

Beale, E. M. L., Kendall, M. G., and Mann, D. W. "The Discarding of Variables in Multivariate Analysis." *Biometrika,* 54 (1967), 3-4.

Bechtel, G. G. "A Dual Scaling Analysis for Paired Compositions." *Psychometrika,* 36 (1971), 135-154.

———. "The Multivariate Analysis of Variance and Multiattribute Scaling." *Oregon Research Institute Bulletin,* 8 (1968), 2.

Bechtel, G. G. and Chaplin, W. "Multidimensional Unfolding of Interval Utility and Similarity Scales." *Oregon Research Institute Bulletin,* 11 (1971), 11.

Bechtel, G. G., Tucker, L., and Chang, W. C. "Linear Multi-dimensional Scaling of Choice," Research Bulletin No. RB-69-73. Educational Testing Service, Princeton, 1969.

Blalock, H. M. Jr. "Estimating Measurement Error Using Multiple Indicators and Several Points in Time." *American Sociological Review,* 35 (1970) 101-111.

———. Causal Inferences in Non-Experimental Research. Durham: University of North Carolina Press, 1964.

———. "Making Causal Inferences for Unmeasured Variables from Correlations among Indicators." *American Journal of Sociology,* 69 (1963), 53-62.

———. "Correlation and Causality: The Multivariate Case." *Social Forces,* 39 (1961), 245-251.

———. "Correlation Analysis and Causal Inferences." *American Anthropologist,* 62 (1960), 624-631.

Bloxom, B. B. "An Alternative Method of Fitting a Model of Individual Differences in Multidimensional Scaling." *Psychometrika,* 39 (1974), 365-367.

———. "Individual Differences in Multi-Dimensional Scaling." Research Bulletin No. RB-68-45, Educational Testing Service, Princeton, 1968.

Bock, R. D. and Bargmann, R. "Analysis of Covariance Structures." *Psychometrika,* 31 (1966), 507-534.

Bock, R. D. and Haggard, E. A. "The Use of Multivariate Analysis of Variance in Behavioral Research." In D. K. Whitla (ed.), *Handbook of Measurement and Assessment in Behavioral Sciences.* Reading, Mass.: Addison-Wesley, 1968.

Boggis, J. G. and Held, I. "Cluster Analysis—A New Tool in Electricity." *Journal of the Market Research Society,* 13 (1971), 49-66.

Bolshev, L. N. "Cluster Analysis." *Proceedings of the International Statistical Institute* (London), 80 (1968), 1-15.

Bonner, R. E. "On Some Clustering Techniques." *IBM Journal of Research and Development,* 8 (1964), 24-34.

Borgnas, L. J. O. "A Study of Attitude Structure Ranking for Automobiles." M.B.A. Thesis, Berkeley, California: Graduate School of Business Administration, University of California at Berkeley, 1969.

Bottenberg, R. and Ward, J. "Applied Multiple Linear Regression Analysis." Technical Report, United States Air Force Personnel Laboratory, Lackland Air Force Base, Texas, 1960.

Boudon, R. "A Method of Linear Causal Analysis: Dependence Analysis." *American Sociological Review,* 30 (1965), 365-374.

Box, G. E. P., and Hill, W. J. "Discrimination Among Mechanistic Models." *Technometrics,* 9 (1967), 57-71.

Boyce, A. J. "Mapping Diversity: A Comparative Study of Some Numerical Methods." in A. J. Cole (ed.), *Numerical Taxonomy.* New York: Academic Press, 1969.

Bradley, J. *Distribution-Free Statistical Tests.* Englewood Cliffs, N.J.: Prentice-Hall, 1968.

Brewer, M. B., Crano, W. D., and Campbell, D. T. "Testing a Single-Factor Model as an Alternative to the Misuse of Partial Correlations in Hypothesis-Testing Research." *Sociometry,* 33 (1970), 1-11.

Browne, M. W. "A Comparison of Factor Analytic Techniques." *Psychometrika,* 33 (1968), 267-334.

Bush, R. R. "Estimation and Evaluation." in R. D. Luce, et al., (eds.), *Handbook of Mathematical Psychology,* Vol. 1. New York: Wiley, 1963.

Calinski, T. "On the Application of Cluster Analysis to Experimental Results." *Proceedings of the International Statistical Institute,* (London), (1969), 108.

Campbell, D. T. "Temporal Changes in Treatment-Effect Correlations: A Quasi-experimental Model for Institutional Records and Longitudinal Studies." *Proceedings of the 1970 Invitational Conference on Testing Problems.* Educational Testing Service, Princeton, 1970.

Carman, J. M., "Correlates of Brand Loyalty: Some Positive Results." *Journal of Marketing Research,* 7 (1970), 67-76.

Carmichael, J. W., George, J. A., and Julius, R. S. "Finding Natural Cluster," *Systematic Zoology,* 17 (1968), 144-150.

Carmone, F. J. "An Investigation of Subjective Evaluation Functions in the Context of University Budgeting." Unpublished Ph.D. thesis, University of Waterloo, 1971.

Carroll, J. D. "Individual Differences and Multidimensional Scaling." In R. N. Shepard, A. K. Romney and S. Nerlove, eds., *Multidimensional Scaling: Theory and Application in the Behavioral Sciences,* Vol. I, New York: Seminar Press, 1972.

―――. "Polynomial Factor Analysis." *Proceedings of the 77th Annual Convention, American Psychological Association* (1969), 103-104.

―――. "Categorical Conjoint Measurement." Ann Arbor, Michigan: Meeting of Mathematical Psychology, August 1969.

―――. "A Generalization of Canonical Correlation Analysis to Three or More Variables." *Proceedings of 76th Annual Convention of the American Psychological Association,* (1968), 227-228.

Cattell, R. B. (ed.) *Handbook of Multivariate Experimental Psychology.* Chicago: Rand McNally, 1966.

―――. "The Three Basic Factor Analytic Research Designs—Their Interrelationships and Derivatives." *Psychological Bulletin,* 19 (1952), 499-520.

―――. "The Scree Test for the Number of Factors," *Multivariate Behavioral Research,* 1 (April, 1966), 245-76.

Chernoff, H. "Metric Considerations in Cluster Analysis." Technical Report No. 67, Department of Statistics, Stanford University, September, 1970.

Christopher, M. G., and Elliott, C. K. "Causal Path Analysis in Market Research." *Journal of the Market Research Society,* 12 (1970), 112-124.

Clevenger, T., Lazier, G. A., and Clark, M. L. "Measurement of Corporate

Images by the Semantic Differential." *Journal of Marketing Research,* 2 (1965), 80-82.

Cochrane, J. L., and Zeleny, M. (eds.) *Multiple Criteria Decision Making.* Columbia, S. C.: University of South Carolina Press, 1973.

Cohen, J. "Multiple Regression as a General Data Analytic System." *Psychological Bulletin,* 70 (1968), 426-443.

Cole, A. J. *Numerical Taxonomy.* New York: Academic Press, 1969.

Cole, J. W. L., and Grizzle, J. E. "Applications of Multivariate Analysis of Variance to Repeated Measurements Experiments." *Biometrics,* 22 (1966), 810-828.

Comack, R. M. "A Review of Classification." *Journal of the Royal Statistical Society,* Series A, 134 (1971), 237-242.

Conger, A. J. and Lipshitz, R. "Canonical Reliability for Profiles and Test Batteries." Report No. 97, L. L. Thurstone Psychometric Laboratory, University of North Carolina, Chapel Hill, 1971.

Cooley, W. W., and Lohnes, P. R. *Multivariate Data Analysis.* New York: Wiley, 1971.

Coombs, C. T. *A Theory of Data.* New York: Wiley, 1964.

―――. "Psychological Scaling Without a Unit of Measurement." *Psychological Review,* 57 (1950), 148-158.

Coombs, C. H., and Kao, R. C. "On a Connection Between Factor Analysis and Multidimensional Unfolding." *Psychometrika,* 25 (1960), 219-231.

―――. "Metric Multidimensional Scaling and the Concept of Preference." Western Management Science Institute, University of California, Working Paper No. 163, Los Angeles, 1970.

Cooper, W. W., and Majone, G. "A Description and Some Suggested Extensions for Methods of Cluster Analysis." Research Report No. 162, Management Science Institute, Carnegie-Mellon University, Pittsburgh, 1968.

Corballis, M. C. "Comparison of Ranks of Cross-Product and Covariance Solutions in Component Analysis." *Psychometrika,* 36 (1971), 243-250.

Cormack, R. M. "A Review of Classification." *Journal of the Royal Statistical Society,* Series A, 134 (1971), 237-242.

Cornfield, J. "Discriminant Functions." *Review of the International Statistical Institute,* 35 (1967), 142-153.

Crask, M., and W. D. Perreault, Jr. "Validation of Discriminant Analysis in Marketing Research," *Journal of Marketing Research,* XIV (February, 1977) 60-68.

Cureton, E. E. "Communality Estimation in Factor Analysis of Small Matrices." *Educational and Psychological Measurement,* 31 (1971), 371-380.

Davidson, J. D. "Forecasting Traffic on STOL." *Operations Research Quarterly,* 24 (1973), 561-569.

Davies, M. G. "The Performance of the Linear Discriminant Function in Two Variables." *British Journal of Mathematical and Statistical Psychology,* 23 (1970), 165-176.

Day, G. S., and Heeler, R. M. "Using Cluster Analysis to Improve Marketing Experiments." *Journal of Marketing Research,* 8 (1971), 340-347.

Dempster, A. P. "An Overview of Multivariate Data Analysis." *Journal of Multivariate Analysis,* 1 (1971), 316-346.

―――. "Stepwise Multivariate Analysis of Variance Based on Principal Variables." *Biometrics,* 19 (1963), 478-490.

Dixon, W. J. (ed.) *Biomedical Computer Programs.* Los Angeles: University of California Press, 1973.

Doehlert, D. H. "Segmenting Subjects and Segmenting Variables in One Analysis." Presented at the First Annual Meeting of the Association for Consumer Research, Amherst, Massachusetts, August, 1970.

Draper, N., and Smith, H. *Applied Regression Analysis.* New York: Wiley, 1966.

Dunn, O. J. "Some Expected Values for Probabilities of Correct Classification in Discriminant Analysis." *Technometrics,* 13 (1971), 345-354.

Eber, H. W. "Toward Oblique Simple Structure: A New Version of Cattell's Maxplane Rotation Program." *Multivariate Behavioral Research,* 1 (1966), 112-125.

Edelstein, R. M., and Rao, V. R. "Subjective Evaluations of Employer Attributes by Administration Students." *Journal of Vocational Behavior,* 6 (1975), 109-120.

Edwards, A. *Experimental Design in Psychological Research.* New York: Holt, Rinehart and Winston, 1960.

Edwards, A. W. F., and Cavalli-Sforza, L. L. "A Method for Cluster Analysis." *Biometrics,* 21 (1965), 362-375.

Ehrenberg, A. S. C. "On Methods: The Factor Analytic Search for Program Types," *Journal of Advertising Research,* 8 (March, 1968), 55-70.

Eisenbeis, R. A., and Avery, R. B. *Discriminant Analysis and Classification Procedures: Theory and Applications.* Lexington, Mass: Lexington Books, 1972.

Ekebald, F. A., and Stasch, S. F. "Criteria in Factor Analysis." *Journal of Advertising Research,* 7 (September, 1967), 48-57.

Ekman, G., and Kuennapas, T. "Distribution Function for Similarity Estimates." *Perceptual and Motor Skills,* 29 (1969), 967-983.

Elton, E. H., and Gruber, M. J. "Homogenous Groups and the Testing of Economic Hypothesis." *Journal of Financial and Quantitative Analysis,* (1970), 581-602.

Estabrook, G. F. "A Mathematical Model in Graph Theory for Biological Classification," *Journal of Theoretical Biology,* 12 (1966), 297-310.

Farley, J. U., and Ring, L. W. "Empirical Specification of a Buyer Behavior Model." *Journal of Marketing Research,* 11 (1974), 89-96.

Farquhar, P. H. and Rao, V.. R. "A Balance Model for Evaluating Subsets of Multiattributed Items." *Management Science,* 22 (January, 1976), 528-539.

Fishbein, M. "Attitude and the Predictors of Behavior." In M. Fishbein (ed.), *Readings in Attitude Theory and Measurement.* New York: Wiley, 1967.

Fisher, R. A. "The Use of Multiple Measurements in Taxonomic Problems." *Annals of Eugenics,* 7 (1936), 179-188.

Fleiss, J. L., and Lubin, J. "On the Methods and Theory of Clustering." *Multivariate Behavioral Research,* 4, 1969, 235-250.

Fortier, J. J. and Solomon, H. "Clustering Procedures." in P. Krishnaiah (ed.), *Multivariate Analysis.* New York: Academic Press, 1966.

Frank, R. E. and Green, P. E. "Numerical Taxonomy in Marketing Analysis: A Review Article," *Journal of Marketing Research,* 5 (1968), 83-94.

Frank, R. E., and Strain, C. E. "A Segmentation Research Design Using Consumer Panel Data." *Journal of Marketing Research,* 9 (1972), 385-390.

Frank, R. E., Massy, W. F., and Morrison, D. G. "Bias in Multiple Discriminant Analysis." *Journal of Marketing Research,* 2 (1965), 250-258.

Freund, R. J., and Early, G. G. "On the Interpretation of the Multivariate Analysis of Variance." Paper presented at the Biometric Society Meetings, May, 1970.

Friedman, H. P., and Rubin, J. "On Some Invariant Criteria for Grouping Data." *Journal of the American Statistical Association,* 62 (1967), 1159-1178.

Frost, W. A. K. "The Development of a Technique for TV Program Assessment." In D. Aaker (ed.) *Multivariate Methods in Marketing: Theory and Applications.* Belmont, California: Wadsworth, 1971.

Gatty, R. "Multivariate Analysis for Marketing Research: An Evaluation." *Applied Statistics,* 15 (1966), 158.

Gebhardt, F. "Maximuum Likelihood Solution to Factor Analysis When Some Factors are Completely Specified." *Psychometrika,* 36 (1971), 155-164.

Gengerelli, J. A. "A Method for Detecting Subgroups in a Population and Specifying Their Membership." *Journal of Psychology,* 55 (1963), 457.

Gibson, W. A. "Three Multivariate Models: Factor Analysis, Latent Structure Analysis, and Latent Profile Analysis." *Psychometrika,* 24 (1959), 229-252.

Gleason, T. C. *Multi-Dimension Scaling of Sociometric Data.* Ann Arbor: Institute for Survey Research, University of Michigan, 1969.

Gnanadesikan, R., and Wilk, M. B. "Data Analytic Methods in Multivariate Statistical Analysis." In P. Krishnaniah (ed.), *Multivariate Analysis,* Vol. 2, New York: Academic Press, 1969.

Goldberger, A. S. "Econometrics and Psychometrics: A Survey of Communalities." *Psychometrika,* 36 (1971), 83-108.

Goldfeld, S. M., and Quandt, R. E. *Nonlinear Methods of Econometrics.* Amsterdam and London: North-Holland, 1972.

Gollob, H. F. "Confounding of Sources of Variation in Factor-Anelytic Techniques." *Psychological Bulletin,* 70 (1968), 330-344.

Goodman, L. A. "The Analysis of Multidimensional Contingency Tables: Stepwise Procedures and Direct Estimation Methods for Building Models for Multiple Classification." *Technometrics,* 13 (1971), 33-62.

Gorman, J. W. and Toman, R. J. "Selection of Variables for Fitting Equations to Data." *Technometrics,* 8 (1966), 27-51.

Gower, J. C., and Hill, I. D. "Internal Data Structures." *Journal of the Royal Statistical Society,* Series C, 20 (1971), 32-44.

Green, B. F. "Latent Structure Analysis and Its Relation to Factor Analysis." *Journal of the American Statistical Association,* 47 (1952), 71-76.

Green, P. E. "On the Design of Choice Experiments Involving Multifactor Alternatives." *Journal of Consumer Research,* 1 (September, 1974) 61-68.

———. "On the Analysis of Interactions in Marketing Research Data." *Journal of Marketing Research,* 10 (1973), 410-420.

———. "Measurement and Data Analysis." *Journal of Marketing,* 34 (1970) 15-17.

Green, P. E., and Tull, D. S. *Research for Marketing Decisions.* Third Edition. Englewood Cliffs, N.J.: Prentice-Hall, 1975.

Green, P. E., and Rao, V. R. *Applied Multidimensional Scaling: A Comparison of Approaches and Algorithms.* New York: Holt, Rinehart & Winston, 1972.

———. "Conjoint Measurement for Quantifying Judgmental Data." *Journal of Marketing Research,* 8 (1971), 355-363.

———. "A Note on Proximity Measures and Cluster Analysis." *Journal of Marketing Research,* 6 (1969), 359-364.

Green, P. E., and Wind, Y. *Multiattribute Decisions in Marketing: A Measurement Approach.* Hinsdale, Ill.: Dryden Press, 1973.

Green, P. E., and Carmone, F. J. *Multidimensional Scaling and Related Techniques.* Boston: Allyn and Bacon, 1970.

————. "Multidimensional Scaling—An Introduction and Comparison of Non-metric Unfolding Techniques." *Journal of Marketing Research,* 6 (1969), 330-341.

Green, P. E., Carmone, F. J., and Wind, Y. "Subjective Evaluation Models and Conjoint Measurement." *Behavioral Science,* 17 (1972), 288-299.

Green, P. E., Carmone, F. J., and Robinson, P. J. "Nonmetric Scaling Methods: An Exposition and Overview." *Wharton Quarterly,* 2 (1968), 159-173.

Green, P. E., Wind, Y., and Jain, A. K. "Analysis of Free Response Data in Marketing Research." *Journal of Marketing Research,* 10 (1973), 45-52.

————. "A Note on the Measurement of Social-Psychological Belief Systems." *Journal of Marketing Research,* 9 (1972), 204-208.

————. "Benefit Bundle Analysis." *Journal of Marketing Research,* 9 (1972), 31-36.

Green, P. E., Frank, R. E., and Robinson, P. J. "Cluster Analysis in Test Market Selection." *Management Science,* 13 (1967), 387-400.

Green, P. E., Halbert, M. W., and Robinson, P. J. "Canonical Analysis—An Exposition and Illustrative Application." *Journal of Marketing,* 3 (1960), 32-39.

Greeno, D. W., Somers, M. S., and Kernan, J. B. "Personality and Implicit Behavior Patterns." *Journal of Marketing Research,* 10 (1973), 63-69.

Guttman, L. "A General Non-metric Technique for Finding the Smallest Coordinate Space for a Configuration of Points." *Psychometrika,* 33 (1968), 496-506.

————. "Best Possible Systematic Estimates of Communality." *Psychometrika,* 21 (1956), 273-285.

————. "Image Theory for the Structure of Quantitative Variates." *Phychometrika,* 18 (1953), 277-296.

————, "A Basis for Scaling Qualitative Data." *American Sociological Review,* 9 (1944), 139-150.

Hair, J. F. "A Tutorial on Canonical Correlation Analysis," in T. Green and S. Jones (eds.), *Applications Issues, Developments, and Strategies in the Decision Sciences,* S. E. American Institute for Decision Sciences, Mississippi State, Mississippi, 1973, 369-372.

Haitovsky, Y. "A Note on the Maximization of R^2." *American Statistician,* 23 1969), 20-21.

Hakastian, A. R. "A Comparative Evaluation of Several Prominent Methods of Oblique Factor Transformation." *Psychometrika,* 36 (1971), 175-194.

Haley, R. I. "Benefit Segmentation: A Decision Oriented Research Tool." *Journal of Marketing,* 32 (1968), 30-35.

Hammarling, S. J. *Latent Roots and Latent Vectors.* Toronto: University of Toronto Press, 1970.

Harman, H. H. *Modern Factor Analysis.* Chicago: The University of Chicago Press, 1967.

Harman, H., and Jones, W. "Factor Analysis by Minimizing Residuals." *Psychometrika,* 31 (1966), 351-368.

Harris, B. (ed.) *Advanced Seminar on Spectral Analysis of Time Series.* New York: Wiley, 1967.

Harris, C. W., and Kaiser, H. F. "Oblique Factor Analysis Solutions by Orthogonal Transformations." *Psychometrika,* 29 (1964), 347-362.

Harris, M. L., and Harris, C. W. "A Factor Analytic Interpretation Strategy." Technical Report No. 115, Wisconsin Research and Development Center for Cognitive Learning, University of Wisconsin, Madison, 1970.

Harris, R. J. *A Primer of Multivariate Statistics,* Academic Press: New York, 1975.

Hays, W. L., *Statistics for Psychologists.* New York: Holt, Rinehart, and Winston, 1963.

Hendrickson, A., and White, P. "PROMAX: A Quick Method for Rotation to Oblique Simple Structure." *British Journal of Statistical Psychology,* 17 (1964), 65-70.

Hoeke, R. S. and R. E. Potter. "Research Using Factor Analysis," in T. Green and S. Jones (eds.), *Applications Issues, Developments, and Strategies in the Decision Sciences,* S. E. American Institute for Decision Sciences, Mississippi State, Mississippi, 1973, 275-278.

Hope, K. "The Complete Analysis of A Data Matrix: Application and Interpretation." *British Journal of Psychiatry,* 116 (1970), 657-666.

──────. *Methods of Multivariate Analysis.* London: University of London Press, 1968.

Horan, C. B. "Multidimensional Scaling: Combining Observations When Individuals Have Different Perceptual structures." *Psychometrika,* 34 (1969), 139-165.

Horst, P. *Factor Analysis of Data Matrices.* New York: Holt, Rinehart and Winston, 1965.

──────. "Generalized Canonical Correlations and Their Applications to Experimental Data." *Journal of Clinical Psychology* (Monograph Supplement), 14 (1961), 331-377.

──────. "Relations Among m Sets of Measures." *Psychometrika,* 26 (1961), 129-149.

Hotelling, H. "Relations Between Two Sets of Variates." *Biometrika,* 28 (1936), 321-377.

──────. "The Most Predictable Criterion." *Journal of Educational Psychology,* 26 (1935), 139-142.

Howard, K. I., and Diesenhaus, H. I. "Direction of Measurement and Profile Similarity." *Multivariate Behavioral Research,* 2 (1967), 225-237.

Hubert, L. "Min and Max Hierarchical Clustering Using Asymmetric Similarity Matrices." *Psychometrika,* 38 (1973), 63-72.

Huberty, C. J. "Multivariate Indices of Strength of Association." *Multivariate, Behavioral Research,* 7 (1972), 523-536.

Huberty, C. J. "Discriminant Analysis," *Review of Educational Research,* 45 (Fall, 1975) 543-598.

Hummel, T. J., and Sligo, J. R. "Empirical Comparisons of Univariate and Multivariate Analysis of Variance Procedures." *Psychological Bulletin.* 76 (1971), 49-57.

Hurst, R. L. "Qualitative Variables in Regression Analysis." *American Educational Research Journal,* 7 (1970), 541-552.

Jardine, N. "Towards a General Theory of Clustering." *Biometrics,* 25 (1969), 609-610.

Jardine, N., and Sibson, R. "The Construction of Hierarchic and Nonhierarchic Classification." *Computer Journal,* 11 (1968), 117-184.

Jenkins, G. M., and Watts, D. G. *Spectral Analysis and Applications.* San Francisco: Holden-Day, 1968.

Jennings, E. "Fixed Effects Analysis of Variance by Regression Analysis." *Multivariate Behavioral Research,* 2 (1967), 95-108.

Johansson, J. K., and Sheth, J. N. "Canonical Correlation and Competitive

Market Structure." Proceedings of the American Institute for Decision Sciences, November, 1973.

――――. "Canonical Correlation, Multiple Regression and Simultaneous Systems." Working Paper, College of Commerce and Business Administration, University of Illinois, Urbana, Illinois, 1973.

Johansson, J. K. and Lewis, C. "A Clarification of the Redundancy Index." Working Paper, College of Commerce and Business Administration, University of Illinois, Urbana, Illinois, 1974.

Johnson, R. L., and Wall, D. D. "Cluster Analysis of Semantic Differential Data." *Psychological Management,* 29 (1969), 769-780.

Johnson, R. M. "Trade-Off Analysis of Consumer Values." *Journal of Marketing Research,* 11 (1974), 121-127.

――――. "Varieties of Conjoint Measurement." Working Paper, Market Facts, Inc., Chicago, 1973.

――――. "Pairwise Nonmetric Multidimensional Scaling." *Psychometrika,* 38 (1973), 11-18.

――――. "Multiplicative Conjoint Measurement." Working Paper, Market Facts, Inc., Chicago, October, 1972.

Johnson, S. C. "Hierarchical Clustering Schemes." *Psychometrika,* 32 (1967), 241-254.

Jollife, I. T. "Discarding Variables in Principal Component Analysis II: Real Data." *Applied Statistician,* 22 (1973), 21-31.

Jones, K. J. *The Multivariate Statistical Analyzer.* Cambridge: Harvard University Bookstore, 1964.

Jones, L. V. "Some Illustrations of Psychological Experiments Designed for Multivariate Analysis." *Psychometric Laboratory,* University of North Carolina, #28, December, 1960.

Joreskog, K. G. "Simultaneous Factor Analysis in Several Populations." Research Bulletin No. RB-70-61, Educational Testing Services, Princeton, (1970).

――――. "Some Contributions to Maximum Likelihood Factor Analysis." *Psychometrika,* 32 (1967), 443-482.

Joreskog, K. G., and Goldberger, A. S. "Factor Analysis by Generalized Least Squares." Research Bulletin No. RB-71-26, Educational Testing Service, Princeton, 1971.

Joreskog, K. G., and Lawley, D. "New Methods in Maximum Likelihood Factor Analysis." Research Bulletin No. RB-67-49, Educational Testing Service, Princeton, 1967.

Kaiser, H. F., Hunka, S., and Bianchini, J. C. "Relating Factors Between Studies Based Upon Different Individuals." *Multivariate Behavior Research,* 6 (1971), 409-422.

Kamen, J. M. "Quick Clustering." *Journal of Marketing Research,* 7 (1970), 199-204.

Kearns, R. J. "Various Concepts About Latent Space: A Survey of Latent Structure Models." Berkeley, California: Center for Research in Management Science, University of California, 1968.

Kelly, L. G. *Handbook of Numerical Methods and Applications.* Reading, Massachusetts: Addison-Wesley, 1967.

Kendall, M. G. "Cluster Analysis." Unpublished manuscript, *Scientific Control Systems, Ltd.,* London, 1968.

――――. *A Course in Multivariate Analysis.* London: Charles Griffin, 1961.

————. "Factor Analysis at a Statistical Technique." *Journal of the Royal Statistical Society,* 12 (1950), 60-73.

Kernan, J. B. "Choice Criteria, Decision Behavior and Personality." *Journal of Markekting Research,* 5 (1968), 155-169.

Kettenring, J. R., "Canonical Analysis of Several Sets of Variables." *Biometrika,* 58 (1971), 433-451.

King, B. F. "Step-wise Clustering Procedures." *Journal of the American Statistical Association,* 62 (1967), 86-101.

————. "Step-wise Cluster Procedures." *Journal of the American Statistical Association,* 62 (1967), 139-169.

Karmer, C. Y., and Jensen, D. R. "Fundamentals of Multivariate Analysis. Part IV. Analysis of Variance for Balanced Experiments." *Journal of Quality Technology,* 2 (1970), 32-40.

————. "Fundamentals of Multivariate Analysis. Part III. Analysis of Variance for One-way Classifications." *Journal of Quality Technology,* 1 (1969), 264-276.

————. "Fundamentals of Multivariate Analysis. Part II. Inference about Two Treatments." *Journal of Quality Technology,* 1 (1969), 189-204.

————. "Fundamentals of Multivariate Analysis. Part I. Inference about Means." *Journal of Quality Technology,* 1 (1969), 120-133.

Kruskal, J. B. "Monotone Regression: Continuity and Differentiability Properties." *Psychometrika,* 36 (1971), 57-62.

————. "How to Use MDSCAL—A Program to do Multidimensional Scaling and Multidimensional Unfolding." Unpublished Manuscript, Bell Telephone Laboratories, 1968.

————. "Nonmetric Multidimensional Scaling: A Numerical Method." *Psychometrika,* 29 (1964), 115-129.

Kruskal, J. B., and Carmone, F. J. "How to Use M-D-Scal (version 5M) and Other Useful Information." Bell Telephone Laboratories, Murray Hill, N.J., March, 1969 (multilithed report).

Kruskal, J. B., and Shepard, R. N. "A Nonmetric Variety of Linear Factor Analysis." *Psychometrika,* 39 (1974), 123-158.

Lachenbruch, P. A. and Mickey, M. R. "Estimation of Error Rates in Discriminant Analysis." *Technometrics,* 10 (1968), 1-11.

Ladd, G. W. "Linear Probability Functions and Discriminant Functions." *Econometrica,* 34 (1966), 873-885.

Lambert, Z., and D. Durand. "Some Precautions in Using Canonical Analysis," *Journal of Marketing Research,* XII (November, 1975) 468-475.

LaMotte, L. R., and Hocking, R. R. "Computational Efficiency in the Selection of Regression Variables." *Technometrics,* 12 (1970), 83-94.

Lance, G., and Williams, W. "A General Theory of Classificatory Sorting Strategies: II. Clustering Systems." *Computer Journal,* 10 (1967), 271.

Lavidge, R. C. and Steiner, G. A. "A Model for Predictive Measurement of Advertising Effectiveness." *Journal of Marketing,* 25 (October, 1961), 59-62.

Lawley, D.. N. "The Application of thhe Maximum Likelihood Method for Factor Analysis." *British Journal of Psychology,* 33 (1943), 172-175.

Lawley, D. N., and Maxwell, A. E. *Factor Analysis as a Statistical Method,* London: Butterworth, 1963.

Lazarsfeld, P. F. "Latent Structure Analysis and Test Theory." In Gulliksen and Messick (eds.), *Psychological Scaling Theory and Applications.* New York: Wiley, 1960.

Lazarsfeld, P. F., and Henry, N. W. *Latent Structure Analysis.* Boston: Houghton Mifflin, 1968.

————. "The Application of Latent Structure Analysis to Quantative Ecological Data." In F. Massarik and P. Ratoosh, (eds.), *Mathematical Explorations in Behavioral Science.* Homewood, Illinois: R. D. Irwin, 1965.

Lehman, D. "Some Alternatives to Linear Factor Analysis for Variable Grouping Applied to Buyer Behavior." *Journal of Marketing Research,* 10 (1974), 206-213.

Lessig, V. P., and Tollefson, J. D. "Market Segmentation Through Numerical Taxonomy." *Journal of Marketing Research,* 8 (1971), 480-487.

Levine, R. L., and Hunter, J. E. "Statistical and Psychometric Inference in Principal Components Analysis." *Multivariate Behavioral Research,* 6 (1971), 105-116.

Lorr, M. et al. "Conference on Cluster Analysis of Multivariate Data, New Orleans, Louisiana, Final Report." Office of Naval Research, AD 653-722, June, 1967.

Luce, R. D., and Tukey, J. W. "Simultaneous Conjoint Measurement: A New Type of Fundamental Measurement." *Journal of Mathematical Psychology,* 1 (1964), 1-27.

Lunney, G. H. "Using Analysis of Variance with a Dichotomous Dependent Variable: An Empirical Study." *Journal of Educational Measurement,* 7 (1970), 263-269.

Lutz, R. J., and Howard, J. A. "Toward a Comprehensive View of the Attitude Behavior Relationship: The Use of Multiple-Set Canonical Analysis." Proceedings, American Statistical Association, Denver, 1971.

MacQueen, J. "Some Methods of Classification and Analysis of Multivariate Observations." In *Fifth Symposium of Mathematical Statistics and Probability,* Berkeley: University of California Press, 1967.

McClain, J. O. and Rao V. R. "Trade-Offs and Conflicts in Evaluation of Health Systems Alternatives: A Methodology for Analysis." Health Services Research, 9 (1974), 35-52.

McCracken, R. F. "Multidimensional Scaling and the Measurement of Consumer Perception: A Description and Sensitivity Analysis." Unpublished MBA thesis, Wharton School, University of Pennsylvania, May, 1968.

McDonald, R. P. "The Theoretical Foundations of Principal Components Factor Analysis, Canonical Factor Analysis and Alpha Factor Analysis." *British Journal of Mathematical and Statistical Psychology,* 23 (1970), 1-21.

————. "A General Approach to Non-linear Factor Analysis." *Psychometrika,* 27 (1962), 203-206.

McGee, V. E. "Nonmetric Multidimensional Scaling—Some Sensitivity Analyses." Presented at the American Marketing Association Fall Conference, Boston, 1970.

————. "Multidimensional Scaling of N Sets of Similarity Measures: A Nonmetric Individual Differences Approach." *Multivariate Behavioral Research,* 3 (1968), 233-248.

McKeon, J. J. "Canonical Analysis: Some Relations Between Canonical Correlation, Factor Analysis, Discriminant Analysis, and Scaling Theory," *Psychometric Monograph,* 13 (1966).

McNeil, K. A. "Meeting the Goals of Research with Multiple Linear Regression." *Multivariate Behavioral Research,* 5 (1970), 375-390.

McQuitty, L. L. "A Comparative Study of Some Selected Methods of Pattern

Analysis." *Educational and Psychological Measurement,* 31 (1971), 607-626.

——. "Hierarchical Classification by Multiple Linkages." *Educational and Psychological Measurement,* 30 (1971), 3-10.

——. "Clusters from Iterative, Intercolumnar Correlation Analysis." *Educational and Psychological Measurement,* 28 (1968), 211-238.

——. "Elementary Linkage Analysis for Isolating Orthogonal and Oblique Types and Typal Relevancies." *Educational and Psychological Measurement,* 17 (1957), 207-229.

McQuitty, L. L., and Clark, J. A., "Clusters from Iterative, Intercolumnar Correlational Analysis." *Educational and Psychological Measurement,* 28 (1968), 211-238.

Marriott, F. H. C. "Practical Problems in a Method of Cluster Analysis." *Biometrics,* 27 (1971), 501-515.

Massy, W. F. "Principal Components Regression in Exploratory Statistical Research." *Journal of the American Statistical Association,* 60 (March, 1965), 234-256.

——. "Applying Factor Analysis to a Specific Marketing Problem." In S. A. Greyson (ed.) *Toward Scientific Marketing.* Chicago: American Marketing Association, (1964), 291-307.

Meredith, W. "Canonical Correlations with Fallible Data?" *Psychometrika,* 29 (1964), 55-66.

Miller, J. K., and Farr, S. D. "Bimultivariate Redundancy: A Comprehensive Measure of Interbattery Relationship." *Multivariate Behavioral Research,* 6 (1971), 313-324.

Miller, R., Eyman, R. K., and Dingman, H. F. "Factor Analysis, Latent Structure Analysis, and Mental Typology." *British Journal of Statistical Psychology,* 52 (1961), 29-33.

Miller, R. G. "Statistical Prediction by Discriminant Analysis." *Meteorlogical Monographs,* 4 (1962), No. 25.

Morrison, D. F. *Multivariate Statistical Methods.* New York: McGraw-Hill, 1967.

Morrison, D. G. "On the Interpretation of Discriminant Analysis." *Journal of Marketing Research,* 6 (1969), 156-163.

——. "Measurement Problems in Cluster Analysis." *Management Science,* 13 (1967), 775-780.

Mosteller, F. "The Jackknife." *Review of the International Statistical Institute,* 39 (1971), 363-372.

Mosteller, F. and Bush, R. R. "Selective Quantitative Techniques." In Gardner Lindzey (ed.), *Handbook of Social Psychology,* Vol. 1. Reading, Mass.: Addison-Wesley.

Mosteller, F., and Tukey, J. W. "Data Analysis, Including Statistics." In G. Lindzey and E. Aronson (eds.), *Handbook of Social Psychology* (2nd edition). Reading, Mass.: Addison-Wesley, Volume 2, 1968.

Myers, J. G., and Nicosia, F. M. "On the Dimensionality Question in Latent Structure Analysis." Paper presented at the Fall Conference of the American Marketing Association, Cincinnati, August, 1969.

——. "On Some Applications of Cluster Analysis for the Study of Consumer Typologies, and Attitudinal-Behavioral Change." In J. Arndt (ed.`, *Insights Into Consumer Behavior.* Boston: Allyn and Bacon, 1968

Nelson, M. J. "Cluster Analysis as a Tool in Revealing Cognitive Clarity." M.B.A. Thesis. Berkeley, California: Graduate School of Business Administration, University of California at Berkeley, 1969.

Nicewander, W. A., and Wood, D. A. "Comments on 'A General Canonical Correlation Index.' " *Psychological Bulletin,* 81 (1974), 92-94.

Nie, N., Bent, D., and Hull, C. *Statistical Package for the Social Sciences.* New York: McGraw-Hill, 1970.

Ostlund, L. "Factor Analysis Applied to Predictors of Innovative Behavior," *Decision Sciences,* 4 (January, 1973), 92-107.

Overall, J. E., and Klett, C. J. *Applied Multivariate Analysis.* New York: McGraw-Hill, 1972.

Parsons, L. J., and Ness, T. E. "Using AID and MCA to Analyze Marketing Data." Presented at the Fall Conference of the American Marketing Association, Minneapolis, 1971.

Parzen, E. "The Role of Spectral Analysis in Time Series Analysis." *Review of the International Statistical Institute,* 35 (1967), 125-141.

Perry, M., and Hamm, B. C. "Canonical Analysis of Relations Between Socio-econometric Risk and Personal Influence in Purchase Decisions." *Journal of Marketing Research,* 6 (1969), 351-354.

Press, L. I., Rogers, M. S., and Shure, G. H. "An Interactive Technique for the Analysis of Multivariate Data." *Behavioral Science,* 14 (1969), 364-370.

Ramond, C. K. "Factor Analysis: When to Use It." In A. Schuchman (ed.) *Scientific Decision Making in Business.* New York: Holt, Rinehart and Winston, 1963.

Rao, C. R. "Estimation and Tests and Significance in Factor Analysis." *Psychometrika,* 20 (1955), 93-111.

Rao, V. R., and Wilcox, G. W. "Multidimensional Psychophysics: An Overview." In Peter Wright (ed.) *Advances in Consumer Research,* Vol. I., Association for Consumer Research, 1973.

Rohlf, J. "Adaptive Hierarchical Clustering Schemes." *Systematic Zoology.* 19 (1970), 58-82.

⸺. "Correlated Characters in Numerical Taxonomy." *Systematic Zoology.* 16 (1967), 109-126.

Roscoe, A. M., Sheth, J. N., and Howell, W. "Intertechique Cross Validation in Cluster Analysis." AMA Educators Conference, Portland, Oregon, 1974. Combined Proceedings. AMA: Chicago, 1974.

Rouanet, H., and Lepine, D. "Comparison Between Treatments in A Repeated-Measurement Design: Anova and Multivariate Methods." *British Journal of Mathematical and Statistical Psychology,* 23 (1970), 147-163.

Roy, S. N., "Stepdown Procedures in Multivariate Analysis." *Annals of Mathematical Statistics,* 29 (1958), 1177-1187.

Rubin, H. "Decision Theoretic Approach to Some Multivariate Problems." In P. Krishnaiah (ed.), *Multivariate Analysis—II.* New York: Academic Press, 1969.

Rulon, P. J., and Brooks, W. D. "On Statistical Tests of Group Differences." In D. K. Whitla (ed.), *Handbook of Measurement and Assessment in Behavioral Sciences.* Reading, Mass.: Addison-Wesley, 1968.

Sammon, J. W. "A Nonlinear Mapping for Data Structure Analysis." *IEEE Computer Journal,* 18 (1966), 401-409.

Scheffe, H. *The Analysis of Variance.* New York: Wiley, 1959.

Schoneman, P. H. "On Metric Multidimensional Unfolding." *Psychometrika*, 35 (1970), 349-366.

Scott, A. J., and Symons, M. J. "Clustering Methods Based on Likelihood Ratio Criteria." *Biometrics*, 27 (1971), 387-398.

Sethi, S. P. "Comparative Cluster Analysis for World Markets." *Journal of Marketing Research*, 8 (1971), 348-354.

Shepard, R. N. "Metric Structures in Ordinal Data." *Journal of Mathematical Psychology*, 3 (1966), 287-315.

Shepard, R. N., and Carroll, J. D. "Parametric Representation of Nonlinear Data Structures." In P. R. Krishnaiah, (ed.), *Multivariate Analysis*, Vol. I, New York: Academic Press, 1966.

Shepard, R. N., Romney, A. K., and Nerlove, S., (eds.) *Multidimensional Scaling Theory and Application in Behavioral Sciences*, Vol. 1. New York: Seminar Press, 1972.

Sheth, J. N. "Multivariate Revolution in Market Research." *Journal of Marketing*, 35 (1971), 13-19.

————. "Canonical Analysis of Attitude-Behavior Relationship." Paper presented at TIMS XVIII International Meeting, Washington, D.C., March 21-24, 1971.

————. "Multivariate Analysis in Marketing." *Journal of Advertising Research*, 10, No. 1, (1970), 29-40.

————. "Measurement of Multidimensional Brand Loyalty of A Consumer." *Journal of Marketing Research*, 7 (1970), 348-354.

————. "Estimating Parameters Using Factor Analysis." *Journal of the American Statistical Association*, 64 (1969), 808-822.

————. "A Factor Analytic Model of Brand Loyalty." *Journal of Marketing Research*, 4 (1968), 395-404.

Sheth, J. N., and Armstrong, J. S. "Factor Analysis of Marketing Data: A Critical Evaluation," AMA Educator's Conference, 1969.

Smith, H. F. "Interpretation of Adjusted Treatment Means and Regressions in Analysis of Covariance." *Biometrics*, 13 (1957), 282-308.

Smith, H., Gnanadeskian, H. and Hughes, J. B. "Multivariate Analysis of Variance (MANOVA)." *Biometrics*, 18 (1962), 22-41.

Smith, Leon, "The Eta Coefficient in MANOVA." *Multivariate Behavioral Research*, 7 (1972), 361-378.

Sneath, P. H. A. "A Comparison of Different Clustering Methods as Applied to Randomly-Spaced Points." *Classification Society Bulletin*, 1 (1965), 2-18.

Somers, R. H. "A Partitioning of Ordinal Information in a Three-Way Cross-classification." *Multivariate Behavioral Research*, 5 (1970), 217-240.

Sonquist, J. A. *Multivariate Model Building*. Institute for Survey Research, University of Michigan, Ann Arbor, 1970.

Sparks, D. L., and Tucker, W. T. "A Multivariate Analysis of Personality and Product Use." *Journal of Marketing Research*, 8 (1971), 67-70.

Spence, I. "Multidimensional Scaling: An Empirical and Theoretical Investigation." Ph.D. Dissertation, University of Toronto, 1970.

Stanton, J. L., and Lowenhar, J. A. "Psychological Need-Product Attribute Congruence as a Basis for Product Planning." Proceedings, AIDS National Meeting, Atlanta, Georgia, 1974.

Stevens, J. P. "Step-down Analysis and Simultaneous Confidence Intervals in MANOVA." *Multivariate Behavioral Research*, 8 (1973), 391-402.

————. "Global Measures of Association in Multivariate Analysis of Variance." *Multivariate Behavioral Research,* 7 (1972), 373-378.

Stewart, D., and Love, W. "A General Canonical Correlation Index." *Psychological Bulletin,* 70 (1968), 160-163.

Stoetzel, J. "A Factor Analysis of the Liquor Preferences of French Consumers." *Journal of Advertising Research,* 1 (December, 1960), 7-11.

Suits, D. B., "The Use of Dummy Variables in Regression Equations." *Journal of the American Statistical Association,* 52 (1957), 548-551.

Tatsuoka, M. M. *Multivariate Analysis: Techniques for Educational and Psychological Research.* New York: Wiley, 1971.

————. *Discriminant Analysis.* Champaign, Illinois: Institute for Personality and Ability Testing, 1970.

————. "Multivariate Analysis." *Review of Educational Research,* 31 (1969), 739-743.

Theil, H. *Principles of Econometrics.* New York: Wiley, 1971.

————. "On the Estimation of Relationships Involving Qualitative Variables." *American Journal of Sociology,* 76 (1970), 103-154.

Thurstone, L. L. *Multiple Factor Analysis.* Chicago: University of Chicago Press, 1947.

Tintner, G. "Some Applications of Multivariate Analysis in Economic Data??" *Journal of the American Statistical Association,* 41 (1946), 472-500.

Togerson, W. S. *Theory and Methods of Scaling.* New York: Wiley, 1965.

————. "Multidimensional Scaling: I—Theory and Method." *Psychometrika,* 17 (1952), 401-419.

Trawinski, I. M., and Bargmann, R. "Maximum Likelihood Estimation with Incomplete Multivariate Data." *Annals of Mathematical Statistics,* 35 (1964), 647-657.

Tryon, R. C. "General Dimensions of Individual Differences: Cluster Analysis Versus Factor Analysis." *Educational and Psychological Measurement,* 18 (1958), 477-495.

————. "Cumulative Communality Cluster Analysis." *Educational and Psychological Measurement,* 18 (1958), 3-35.

Tryon, R. C., and Bailey, D. E. *Cluster and Factor Analysis.* Mimeo. Department of Psychology, University of California, Berkeley.

————. *Cluster Analysis.* New York: McGraw-Hill, 1971.

Tryon, R. C. "General Dimensions of Individual Differences: Cluster Analysis versus Factor Analysis." *Educational and Psychological Measurement,* 18 (1958), 477-495.

Tucker, L. R. "Topics in Factor Analysis II." AD 717-680, University of Illinois, Urbana, 1970.

Twedt, D. W. "A Multiple Factor Analysis of Advertising Readership." *Journal of Applied Psychology,* 36 (1952), 207-215.

Urbankh, V. Y., "A Discriminant Method of Clustering." *Journal of Multivariate Analysis,* 2 (1972), 249-260.

Van de Geer, J. P. *Introduction to Multivariate Analysis for the Social Sciences.* San Francisco: Freeman, 1971.

Veldman, D. *FORTRAN Programming for the Behavioral Sciences.* New York: Holt, Rinehart and Winston, 1967.

Wackwitz, J. H., and Horn, J. L. "On Obtaining the Best Estimates of Factor Scores Within an Ideal Simple Structure." *Multivariate Behavioral Research,* 6 (1971), 389-408.

Waern, Y. "A Comparison of Some Models for Multidimensional Similarity." Psychological Laboratory, University of Stockholm, Report No. 312, 1970.

Wampler, R. H. "A Report on the Accuracy of Some Widely Used Least Squares Computer Programs." *Journal of the American Statistical Association*, 65 (1970), 549-565.

Wang, M. D., and Stanley, J. C. "Differential Weighting: A Review of Methods and Empirical Studies." *Review of Educational Research*, 65 (1971), 663-705.

Ward, J. H. "Hierarchical Grouping to Optimize an Objective Function." *Journal of the American Statistical Association*, 58 (1963), 236-244.

Warren, W. G. "Correlation or Regression: Bias or Precision." *Applied Statistics*, 20 (1971), 148-164.

Waugh, F. W. "Regression Between Sets of Variates." *Econometrica*, 10 (1942), 299-310.

Wells, W. D. "Backward Segmentation." In J. Arndt (ed.) *Insights Into Consumer Behavior*. New York: Allyn and Bacon, 1968.

Wells, W. D., and Sheth, J. N. "Factor Analysis in Marketing Research." In R. Ferber (ed.), *Handbook of Marketing Research*. New York: McGraw-Hill, 1974.

Wind, Y., and Denny, J. "Multivariate Analysis of Variance in Research on the Effectiveness of T. V. Commercials." *Journal of Marketing Research* 11 (1974), 136-142.

Winer, B. J. *Statistical Principles in Experimental Design*. New York: McGraw-Hill, 1971.

Wolfe, J. H. "Pattern Clustering By Multivariate Clustering Analysis." Research Memorandum SRM 69-17, United States Naval Personnel Research Activity, San Diego, 1969.

Wuhlbert, R. "Multivariate Analysis of Dichotomous Variables: A General Method." *Multivariate Behavioral Research*, 6 (1971), 215-232.

Yee, A. H. "The Source and Direction of Causal Influence in Teacher-Pupil Relations." *Journal of Educational Psychology*, 59 (1968).

Yee, A. H., and Gage, N. L. "Techniques for Estimating the Source and Direction of Causal Influence in Panel Data." *Psychological Bulletin*, 70 (1968), 115-126.

Young, F. W. "Conjoint Scaling." Research Paper No. 118, L. L. Thurstone Psychometric Laboratory, University of North Carolina, April, 1973.

———. "Non-metric Multi-dimensional Scaling: Recovery of Metric Information." *Psychometrika*, 35 (1970), 475-492.

———. "TORSCA, an IBM Program for Nonmetric Multidimensional Scaling." *Journal of Marketing Research*, 5 (1968), 319-321.

Young, F. W., and Appelbaum, N. I. "Non-metric Multidimensional scaling: The Relationship of Several Methods." Report No. 71, L. L. Thurstone Psychometric Laboratory, University of North Carolina, Chapel Hill, 1969.

Zellner, A. *An Introduction to Bayesian Inference in Econometrics*. New York: Wiley, 1971.

Ziff, R. "Psychographics for Market Segmentation." *Journal of Advertising Research*, 11 (April, 1971), 3-10.

INDEX

NOTES

NOTES

NOTES

NOTES

NOTES

NOTES